Color Tab Index

Birds are grouped by color. Within a color group they are arranged roughly by size, from the smallest to the largest.

Birds with large amounts of [obscured by barcode]

Birds that have black or strong gray and orange

Birds that are mostly red

Birds that are mostly brown

Birds with large amounts of blue

Birds with any green

Birds that are all white

Birds that are mostly black

Birds that are mostly black and white

Birds that are red, black, and white

Birds that are mostly gray

Stokes Nature Guides

by Donald Stokes

A Guide to Nature in Winter
A Guide to Observing Insect Lives
A Guide to Bird Behavior, Volume I

by Donald and Lillian Stokes

A Guide to Bird Behavior, Volumes II and III
A Guide to Enjoying Wildflowers
A Guide to Animal Tracking and Behavior

by Thomas F. Tyning

A Guide to Amphibians and Reptiles

Stokes Backyard Nature Books

by Donald and Lillian Stokes

The Bird Feeder Book
The Hummingbird Book
The Complete Birdhouse Book
The Bluebird Book
The Wildflower Book — East of the Rockies
The Wildflower Book — From the Rockies West

by Donald and Lillian Stokes/Ernest Williams

The Butterfly Book

Stokes Field Guides

by Donald and Lillian Stokes

Stokes Field Guide to Birds: Eastern Region
Stokes Field Guide to Birds: Western Region

Stokes Beginner's Guides

by Donald and Lillian Stokes

Stokes Beginner's Guide to Birds: Eastern Region
Stokes Beginner's Guide to Birds: Western Region

By Donald Stokes

The Natural History of Wild Shrubs and Vines

Eastern Region

STOKES
Beginner's Guide
to Birds

Donald and Lillian Stokes

Little, Brown and Company

Boston New York Toronto London

First Edition

10 9 8 7 6 5 4 3 2 1

RRD-OH

Designed by Barbara Werden

Published simultaneously in Canada by Little, Brown & Company (Canada) Limited

Printed in the United States of America

Library of Congress Cataloging-in-Publication Data

Stokes, Donald W.
    Stokes beginner's guide to birds. Eastern region / Donald and Lillian Stokes. —
1st ed.
        p.   cm.
    Includes index.
    ISBN 0-316-81811-9
    1. Birds — East (U.S.) — Identification.
I. Stokes, Lillian Q. II. Title.
QL683.E27S756   1996
598.2974 — dc20   96-13002

Contents

How to Use This Guide

The Stokes Beginner's Guide to Birds: Eastern Region includes 100 of the species you are most likely to see when you first start to watch birds. Below is a description of the main parts of the guide and how to use them.

Alphabetical Index

Inside the front and back covers is the alphabetical index. This helps you look up a species you are already familiar with but about which you want more information. The birds are arranged in alphabetical order by their last names. Thus, American Robin is listed under *R* for Robin, Herring Gull is listed under *G* for Gull.

Color Tab Index

The Color Tab Index makes it easy to look up birds.

When you see a bird, determine its main color and use the tab with that color to get you to the right section of the guide. Once there, look among the birds for the one that looks most like the one you have seen. People see colors differently, so if you do not find your bird in your first choice of color tab, try the color that would be your second-best guess. The birds in each section are arranged roughly by size, with the smaller at the start and the larger at the end.

When male and female of a species are different colors, they are shown under different color tabs with a small photograph of the opposite sex inserted in the corner of the main photograph to show that they are the same species.

Identification Pages

Each species identification account starts with the bird's common name in large letters, followed by its scientific name in smaller italic letters. Then the length of the bird is given, measured from the tip of its bill to the tip of its tail. Each species account also includes the following:

I.D. — This section points out the main features of the bird that distinguish it from other species.

In some species, the male and

female look different; in other species, the birds have different plumage in summer and winter. In these cases, you will see the headings MALE, FEMALE, SUMMER, WINTER before the descriptions of the different plumages.

For some species you will see an immature plumage. This is marked in the text by the word IMMATURE and refers to any plumage of a young bird (from its first winter on) that is unlike that of the adult. In some species, such as gulls, young birds take up to 4 years to look like the adults.

For a few species we show the juvenal plumage. This is the plumage worn in the first summer. It is marked by the word JUVENILE.

Voice — This is a description of the main sounds that you may hear the bird make. Knowing these sounds can help you identify birds and also introduce you to their language. Many of our smaller birds give *songs;* these are complex sounds that are partly instinctive and partly learned. In many species, only the male sings. All other vocal sounds birds make are called *calls;* these are usually short, simple sounds that are instinctively given without any previous learning.

Habitat — This describes the areas with their types of plants where you are most likely to find the bird. Each species has its own habitat needs. For example, for nesting, meadowlarks need fields, nuthatches need woods, and mockingbirds need shrubs. Without these habitats, these birds cannot survive. This is why preserving habitats is so important to conserving birds.

This section also shows what has happened to the population of this species from 1966 to 1993. This information is given next to the word POPULATION. It is based on a North American census called the Breeding Bird Survey, in which volunteers go out on a fixed route each year at approximately the same time and count all breeding birds.

If the population has risen less than 40% over the 27-year period, there is a single arrow up (↑); if the population has risen over 40%, there is a double arrow up (⇑). If the population has fallen less than 40%, there is a single arrow down (↓); if the population has fallen more than 40%, there is a double arrow down (⇓).

Things that cause bird populations to decline include loss of nesting or feeding habitat, disease, harsh weather, and pollution. Things that allow bird populations to increase include new feeding or breeding habitat, favorable weather, and spreading out to live in a larger geographical range.

Becoming aware of the population trends of birds is a good way to see which species are in trouble and need our attention and help to survive.

Nesting — Here you will learn all about the fascinating nesting habits of birds. Not only are these interesting to know, they may also help you understand a bird's behavior as you watch it during the breeding period.

First there is a description of the materials and placement of the nest.

Following this are details of the nesting cycle, with letters used to indicate each part of the cycle. These include:

Eggs = The average number of eggs and their color.
I = The number of days spent *incubating* the eggs; this is the time when generally just the female sits over the eggs to warm them so that the chicks can develop inside.
N = The time that the hatched young remain in the nest, called the *nestling phase*. If there is no N, then the young leave the nest upon hatching.
F = The time from when the eggs hatch to when the young can first fly, a moment called *fledging*.
B = The number of times per year

an individual bird is likely to go through a complete nesting cycle, or *brood*.

Attracting — This section indicates what you can do to help provide for a bird's feeding and nesting needs. There are sections on bird feeders and which seeds are favored, birdhouses and their correct dimensions, and any plantings you can add to make your yard a more attractive habitat for that species. Bird feeder, birdhouse, and birdbath symbols quickly point out which species use them.

Dimensions for birdhouses include the diameter of the entrance hole and the distance between the bottom of the hole and the floor of the house. We also mention the size of the floor area.

Range Map — This map gives you a good idea of where a bird lives in summer and winter.

Yellow = Summer range
Green = Year-round range
Blue = Winter range

Use the range maps to help you know if a bird is likely to occur in your area in a particular season. When you are trying to identify a bird, range maps can also help you eliminate certain species from consideration by showing that they do not live in your area.

If the range map for a bird shows only a yellow summer range, this means that the species migrates south for the winter, usually to Central or South America, both of which are outside the area of the map.

Tips on Identifying Birds

Identifying birds is a wonderful challenge and can develop into an exciting skill and hobby that you can enjoy all of your life.

What follows are a few tips for the beginner on how to more successfully identify the birds you see:

1. Describe before you identify. Notice and describe what you see to yourself and try to memorize this before you look up the bird. What colors do you see? Where are the colors on the bird? Where is the bird — on the ground, at a feeder, near water? What is the bird doing — pecking at bark, eating a seed, creeping down a tree?

2. Use the photographs as guidelines for identification. Within a species, individuals vary; some may be slightly darker, some may be slightly larger, some may have slightly longer bills. Even the same individual bird can look different; in cold weather birds will often puff out their feathers to stay warm and in warm weather they may sleek their feathers to cool off.

3. If possible, get binoculars so that you can see birds up close.

When you have learned most of the birds in this guide, go get our more complete field guide, *Stokes Field Guide to Birds: Eastern Region* or *Western Region*.

Attracting Birds

There are four main ways that you can attract more birds to your property: bird feeders, birdhouses, birdbaths, and plantings.

Setting Up Bird Feeders

Feeding birds is a wonderful way to increase the number of birds around your house and to see them up close. Over 65 million people in the United States feed birds in their backyards, and there is a good reason why this is such a popular hobby — it is great fun, it helps the birds, it reduces stress, and it brings you closer to nature.

You can enjoy feeding birds all year; in fact, in some cases you may attract more birds in spring through fall than you will in winter. This is because more species are here in summer than in winter.

There are many kinds of bird feeders: some for seed, some for suet, and others for sugar solutions. Here is a brief description of each type and how to use it:

Seed Feeders — Some feeders are designed for just thistle seed. They are usually tubular, with tiny openings just big enough for the thistle seed. They attract goldfinches, House Finches, Pine Siskins, chickadees, and others.

The vast majority of seed feeders take other types of seed. They may be tubular with perches, or like a small hopper with a platform underneath. Although any kind of seed can be put in these feeders, most birds that come to them prefer just sunflower seed, or a mixed seed with a high proportion of sunflower seed.

Seed feeders can be hung or mounted on a pole. In either case, place them 8 ft. from the nearest place from which a squirrel could jump, and try to get a baffle — a disklike shield placed above or below a feeder — to help keep the squirrels off. Birds that regularly come to these types of feeders include chickadees, titmice, nuthatches, House Finches, goldfinches, jays, and others.

You can also scatter seed directly on open ground underneath your other feeders or on a small traylike platform just above the ground. This is good for the many species that prefer to feed on the ground, such as cardinals, juncos, sparrows, towhees, Mourning Doves, and others. In these situations try sunflower seed, seed mixes, white millet, or cracked corn.

Suet Feeders — Suet is fat from around the kidneys of cattle. You can buy it at your supermarket meat counter, or you can buy suet cakes from your local bird-feeder supply store. Suet cakes are rendered suet that stays solid better in warm weather. They often contain other things birds like, such as peanuts, fruit, or insect parts. Suet cakes usually fit directly into a wire basket that can be nailed to a tree or hung from a seed feeder.

Birds that are attracted to suet include woodpeckers, chickadees, titmice, nuthatches, and jays.

Hummingbird Feeders — Hummingbirds are easy to attract with special feeders that hold a sugar solution.

To make the sugar solution, combine 1 cup sugar (*not* honey) with 4 cups water and boil for 1–2 minutes. Let cool and then place some in the feeder and store the rest in the refrigerator for later use. Do not add red dye to the solution in an attempt to further attract hummingbirds; it is not needed and may harm the birds. Besides, there is usually red on the feeder.

Be sure to replace the solution every 3 days to keep it fresh and free of mold, which might hurt the birds.

Using Birdhouses

In addition to feeders, a great way to attract birds is by putting up birdhouses. They not only provide needed cavities for nesting, but they also enable you to watch a bird throughout its breeding cycle.

Some species of birds typically nest in tree holes, either a natural cavity or one they excavate. These birds often accept human-made houses that imitate their natural nest holes.

About 25 species of birds commonly use birdhouses. These include chickadees, titmice, wrens, bluebirds, Purple Martins, nuthatches, starlings, some flycatchers, some swallows, some owls, and some woodpeckers. The birds in this guide that use birdhouses have a small symbol of a birdhouse under the section on **Attracting**.

Features of a Good Birdhouse —
There are many types of birdhouses available for sale. Not all are good for birds. Here are a few things to look for in a good birdhouse:

1. It should have only one entrance hole (except in the case of Purple Martin birdhouses).
2. You should be able to open it, so you can monitor the bird's success through nesting and clean out the nest when the young have fledged.
3. It should be the right dimensions for the species you are trying to attract.
4. It should have ventilation holes at the top and drainage holes in the bottom.
5. It should be made of wood, for this insulates the house from heat and cold.
6. It should not have a perch in front of the hole; the birds do not need it, and it might help predators get inside.
7. The roof should overhang the entrance hole to protect it from sun and rain.
8. There should be a way to mount the box to a post or pole.

These are minimum good features. Any birdhouse without these should not be used.

Where and When to Put Up a Birdhouse — Birdhouses can be placed on poles in the open, on trees, on fence posts, or on the sides of buildings. Some birds will tolerate nesting closer to human activity than others.

Most birdhouses should be placed about 5–6 ft. high. This way you can easily clean them out at the end of the breeding season. It does not matter which compass direction they face.

You can place several birdhouses in different locations on your property and see which ones the birds prefer.

You can put up birdhouses at any time of year, but fall through early spring is the best time, for the birds use them in late spring through summer.

Birdhouse Dimensions — Several dimensions of a birdhouse are critical to the safe and successful breeding of the birds that use it:

> *Size of the entrance hole* — The entrance hole must be large and smooth enough to enable the birds to go in and out without too much wear on their feathers.
>
> *Distance from the hole to the floor* — The floor must be far enough beneath the entrance hole to allow the bird room to build its nest and also to prevent predators, such as raccoons, from easily reaching in and grabbing the birds.
>
> *Interior dimensions of the floor* — These must be large enough to accommodate the nest, but not so large that the bird cannot fill the floor area with its nest.

Birdbaths

Water is a major need of birds. They use it for drinking and bathing. If you put out a birdbath, you will not only help the species that you have attracted with feeders, you will also attract other species that do not come to feeders but still love to bathe and drink from birdbaths.

Place a birdbath somewhere near a shrub or tree where the birds can land before coming to the water and where they can dry off and preen after drinking or bathing.

Some people even provide water all winter in colder climates by getting birdbath heaters.

Plantings

You can make your yard more attractive to birds by adding plants that provide food and nesting habitats.

In general, the more varied the plantings in your yard, the more species you are likely to attract. Here are some types of plants you can add and their benefits to the birds:

> *Evergreens* — Plant several evergreen trees together and you will provide shelter for birds in bad weather and at night when they roost. In addition, many species of common birds nest in evergreens, and the cones may attract seed-eating finches in winter.
>
> *Shrubs* — Areas of shrubs are a favorite habitat for many birds. They provide nesting habitats for species such as mockingbirds and thrashers, roosting spots for sparrows, and seeds and berries for a wide variety of small birds.

ediately fol-
er refer to the
ph on the
m; L = left; R =
re inset within

4, 65, 69, 73,

53.

Brown: 3L, 3R,
2, 52, 56i, 66L,
, 108.

Ornithology:
F. K. Truslow —

5, 37, 68, 117.

0, 45i, 81, 91,

4.

Steven D. Faccio: 25.

Sam Fried: 87, 96iR.

Gary Froehlich: 63i, 88, 90, 119iL,
119iR, 121.

James Hill, III: 84.

Kevin T. Karlson: 22, 60.

Stephen G. Maka: 114.

Maslowski Photograph: 4R, 23, 26,
30BR, 34L, 47, 57, 71, 94, 101.

Anthony Mercieca: 75, 97, 103.

Arthur Morris, Birds as Art: 4L, 5, 9,
14, 31, 41, 70, 90i, 99, 123iT.

Photo/Nats: S. G. Maka — 96iL.

Brian Small: 15, 21, 46, 48, 64, 82.

Hugh P. Smith, Jr.: 16, 27, 30L, 50i,
110, 115.

Lillian Stokes: 28, 62, 63, 67, 76, 77,
77i, 78L, 78BR, 78TR, 79, 80, 89,
91i, 92, 93, 95, 95i, 100, 113, 119,
121iT, 121iB, 122, 122iT, 122iB,
123, 123iB, 124, 125, 126.

John Tveten: 61i, 85, 86, 116R.

Tom Vezo: 52R, 55, 59T, 59B, 68i, 1(

VIREO: A. and S. Carey — 44;
W. Greene — 10i, 72L, 104L, 105
K. T. Karlson — 72R; S. J. Lang —
98, 102R; A. Morris — 11, 19, 33l
96, 120i; O. S. Pettingill, Jr. — 17
F. K. Schleicher — 111, 118R; B.
Schorre — 6, 7, 8, 12, 13, 24, 38,
39, 49; J. Schumache — 40; T. J.
Ulrich — 29, 36, 102L; R. Villani
30TR; D. Wechsler — 116L; B. K.
Wheeler — 45, 56.

Brian K. Wheeler: 50, 97i.

Try to plant a variety of
shrubs, some that produce
berries at different seasons.
Grasses and Wildflowers — These
low plants attract insects in
summer and produce seeds
for the winter, and insects
and seeds are important foods
for many common birds. If
you have an area of your
property that you can just let
grow, you will find many
birds attracted to it.

For More Information

For more information on feeders
and attracting birds, see our other
books:

The Bird Feeder Book
The Hummingbird Book
The Complete Birdhouse Book
The Bluebird Book

About Binoculars

Seeing birds up close is a thrilling experience, and binoculars can be a wonderful aid to this. Here are a few tips about binoculars that will help you get a pair that works for you at the right price and the right quality.

Two Important Numbers

All binoculars have 2 basic numbers associated with them, numbers such as 7 x 35 or 8.5 x 42, for example.

The first number refers to the power of the binoculars to magnify an object, by 7 times or by 8.5 times in the examples above. It is best to get binoculars with magnification power somewhere from 7 to 8.5. More powerful than this and you will have trouble holding the image steady as you look through; any less

powerful and you will not see the birds closely enough to identify.

The second number refers to the size, in millimeters, of the opening at the far end of the binoculars. The larger the opening, the more light is let through and the clearer you will see the bird. An opening size somewhere between 35 and 42 is good for most purposes.

Size and Weight

One of the mistakes most beginners make is making weight and size their main criteria for choosing binoculars. They choose very light and very small binoculars so that they can easily take them on hikes and in their pockets.

Unfortunately, very light and

small binoculars rarely have the power or let in enough light to be good for bird-watching. A good pair of bird-watching binoculars will probably not fit in your pocket and will weigh about 20–28 oz.

Buy binoculars at a store that carries several brands so that you can try all of them out. Get advice from other people who have binoculars and try looking through different pairs.

If you enjoy bird-watching, it is worthwhile investing in good binoculars. They will bring the birds closer for the rest of your life, resulting in a tremendous amount of beauty and joy.

Helping to Save the Birds

The most important step we can take to help save our wonderful variety of beautiful birds today is to save the variety of habitats in which they live. If you take the home away from humans, they have trouble surviving. The same is true of birds; the habitats in which they live are their homes.

Ten Ways You Can Help Save Birds

Here are a few actions you can take that will help save birds:

1. Learn to identify more birds.
2. Learn more about bird behavior.
3. Learn the population trends of the birds you see.
4. Create good bird habitats in

your own backyard ing bird feeders, bir birdbaths, and plant attract birds.
5. Keep a notebook of y observations.
6. Share your love for an knowledge of birds wi ers, both young and ol
7. Participate in bird cens and surveys.
8. Join birding organizatio and bird clubs in your a
9. Join local, national, and national conservation or zations.
10. Help your local conservat commission acquire and r age town lands so that the support more bird life.

Identification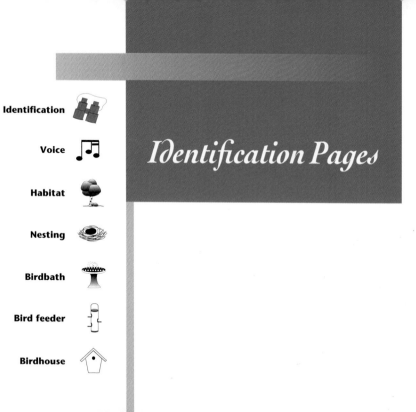

Voice

Habitat

Nesting

Birdbath

Bird feeder

Birdhouse

Identification Pages

American Goldfinch

Carduelis tristis 5"

I.D. **Summer:** Yellow body; black wings and tail. **Male:** Black cap. **Female:** All-yellow head.

Voice Flight call is "perchicoree perchicoree"; song is a long canarylike warble or a short forceful warble.

Habitat Open areas with shrubs and trees, farms, suburban yards, gardens. **Population:** ↓

Nesting Nest of weed bark fastened with caterpillar webbing, placed in shrub or tree. Eggs: 3–7, light blue; I: 12–14 days; F: 11–15 days; B: 1–2.

Attracting

Prefers thistle or hulled sunflower seed in hanging feeders.

Male, summer

Female, summer

Winter, p. 110

Tip Can look like different birds in winter because they change from yellow to mostly grayish brown.

Comes to birdbaths for drinking and bathing.

Yellow Warbler
Dendroica petechia 5"

I.D.

MALE: Yellow with reddish streaks on breast. **FEMALE:** Paler yellow, with breast streaks fainter or absent.

Voice

Song sounds like "sweet sweet sweet, I'm so sweet"; call is a musical "chip."

Habitat

Shrubby areas, especially near water with willows and alder; also yards, gardens.
POPULATION: ↑

Nesting

Nest of milkweed stem fibers, grasses, and down, placed in an upright fork of shrub or small tree. Eggs: 4–6, white with blotches; I: 10 days; F: 9–11 days; B: 1.

Attracting
Planting shrubs near a lake or pond may attract Yellow Warblers to nest.

Male / Female

 Tip Watch for females collecting downy fibers and webbing from tent caterpillar nests to build nests; listen for males singing from exposed perches.

Common Yellowthroat

Geothlypis trichas 5"

I.D.
MALE: Yellow throat and upper breast; black mask with a grayish-white border.

Voice
Song sounds like "your money, your money, your money"; phrase goes with the robberlike mask of the male. Call is a sharp "tchet."

Habitat
Dense brushy habitats near wet areas; also drier habitats with dense understory. POPULATION: ↓

Nesting
Cuplike nest of grasses, leaves, and hair, placed in shrubbery. Eggs: 3–4, creamy white with brown marks; I: 12 days; F: 8–9 days; B: 1–2.

Attracting
Planting shrubs near a lake or pond may attract them to nest.

Female, p. 21

Male

Tip If you get near their nest, the birds will approach and give their "tchet" call.

Orchard Oriole
Icterus spurius 7"

I.D.
FEMALE: Olive-green above; yellow below; 2 thin white wing bars.

Voice
Song is a rapid series of whistled notes; call is a short "chuk."

Habitat
Orchards, open woods, shade trees in towns, parks, streamside groves. **POPULATION:** ↓

Nesting
Pouchlike nest of woven grasses, suspended from tree branch. Eggs: 3–7, pale grayish blue with dark marks; I: 12–14 days; F: 11–14 days; B: 1.

Attracting

Orchard Orioles may come to hummingbird feeders to sip the sugar solution.

Male, p. 12

Female

Tip In spring, may be seen drinking nectar from tree flowers, such as apple blossoms.

Baltimore Oriole

Icterus galbula 8¹/₂"

I.D. **FEMALE:** Orange-yellow below; head and back mottled with black; throat variably mottled with black.

Voice Song is 4–8 medium-pitched whistled notes; calls include a 2-note "teetoo" and a rapid chatter, like "ch'ch'ch'ch."

Habitat Deciduous trees near openings, such as gardens, parks, roads.
POPULATION: ↓

Nesting Nest is a pouchlike structure suspended from tip of tree limb. Woven with plant stem fibers and string. Eggs: 4–6, pale bluish white with dark marks; I: 12–14 days; F: 12–14 days; B: 1.

Female

Male, p. 13

Tip Orioles are obvious in spring, when both males and females repeatedly sing as part of their courtship.

Attracting
May come to sugar solutions in hummingbird or oriole feeders. May also come to orange halves placed out near feeders.

7

Scarlet Tanager

Piranga olivacea 7"

I.D. FEMALE: Yellowish below; greenish above; wings dark grayish brown.

Voice Call is a distinctive "chip-burr."

Habitat Mature deciduous forests.
POPULATION: ↑

Nesting Nest of twigs, grasses, and rootlets, placed on horizontal limb of tree. Eggs: 2–5, pale bluish green with irregular dots of brown; I: 12–14 days; F: 9–10 days; B: 1.

Attracting

Comes to birdbaths for drinking and bathing.

Male, p. 18

Female

Tip **Best discovered by hearing the "chip-burr" call. Female does all nest building and incubating of eggs.**

Summer Tanager

Piranga rubra 7½"

I.D.
FEMALE: Yellowish below; slightly darker above.

Voice
Call is a rapid harsh "chchbit."

Habitat
Pine-oak woods, willows and cottonwoods along streams.
POPULATION: ↓

Nesting
Nest of weed stems, bark, and coarse grasses, placed on horizontal branch of tree. Eggs: 3–5, pale bluish green with dark marks; I: 12 days; F: unknown; B: unknown.

Attracting

Sometimes comes to feeders for fruit, bread crumbs, and suet mixtures.

Comes to birdbaths for drinking and bathing.

Since it is fond of berries, planting berry-producing shrubs will make a more favorable habitat for it.

Male, p. 19

Female

Tip Best found by hearing its call. May fly out to catch insects in midair.

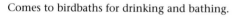

Female

Male

Evening Grosbeak
Coccothraustes vespertinus 8"

I.D.

MALE: Yellow body; black-and-white wings; darker head with a bright yellow eyebrow. FEMALE: Brownish gray overall; black-and-white wings; yellow on back of neck.

Voice

Song is a halting warble; call is a ringing "peer" and when given by a flock is reminiscent of sleighbells.

Habitat

Summers in northern woods; winters in open areas with trees and shrubs. POPULATION: ⇑

Nesting

Nest of twigs and moss, placed at end of tree branch. Eggs: 2–5, blue or bluish green with dark marks; I: 11–14 days; F: 13–14 days; B: 2.

Attracting
Prefers sunflower seed.

Tip **Evening Grosbeaks are seen mostly in winter, when they move south from their northern range.**

Planting trees such as ash, maple, and tulip poplar provides natural seeds for them to eat.

10

Eastern Meadowlark
Sturnella magna 9"

I.D. Streaked brown above; striking yellow below; broad black V on breast. **IN FLIGHT:** Note white outer tail feathers as bird flutters and glides.

Voice Song is 2–8 high-pitched whistles, like "seeoo seeyeer"; call is a "dzeert."

Habitat Meadows and grasslands. **POPULATION:** ⇓

Nesting Nest is a domed structure of grasses, placed in natural or scraped depression in ground. Eggs: 3–7, white with dark marks; I: 13–15 days; F: 11–12 days; B: 1–2.

Tip Most often seen in among field grasses, but males often fly up to top of fence posts or small shrubs to sing.

Attracting
Meadowlarks are dependent for their breeding on large grassy fields. It is important that these fields not be mowed until the young can fly.

Orchard Oriole
Icterus spurius 7"

I.D.
MALE: Black hood, wings, and tail; orange-brown belly and rump.

Voice
Song is a rapid series of whistled notes; call is a short "chuk."

Habitat
Orchards, open woods, shade trees in towns, parks, streamside groves. **POPULATION:** ↓

Nesting
Pouchlike nest of woven grasses, suspended from tree branch. Eggs: 3–7, pale grayish blue with dark marks; I: 12–14 days; F: 11–14 days; B: 1.

Attracting
Orchard Orioles may come to hummingbird feeders to sip the sugar solution.

Male

Female, p. 6

Tip In spring, may be seen drinking nectar from tree flowers such as apple blossoms.

12

Baltimore Oriole
Icterus galbula 8½"

I.D.

MALE: Black hood and back; orange body.

Voice

Song is 4–8 medium-pitched whistled notes; calls include a 2-note "teetoo" and a rapid chatter, like "ch'ch'ch'ch."

Habitat

Deciduous trees near openings, such as gardens, parks, roads.
POPULATION: ↓

Nesting

Nest is a pouchlike structure suspended from tip of tree limb. Woven with plant stem fibers and string. Eggs: 4–6, pale bluish white with dark marks; I: 12–14 days; F: 12–14 days; B: 1.

Male

Female, p. 7

Tip Orioles are obvious in spring, when both males and females repeatedly sing as part of their courtship.

Attracting
May come to sugar solutions in hummingbird or oriole feeders. May also come to orange halves placed out near feeders.

Eastern Towhee
Pipilo erythrophthalmus 8"

I.D.

MALE: Black hood and back; sides orange-brown; belly white.

Voice

Song is a loud "Drink your tea"; call is a loud "chewink."

Habitat

Shrubby edges or open woods with shrub understory.
POPULATION: ⇓

Nesting

Nest of leaves, bark, and grass, placed on or near ground in scratched depression under brush. Eggs: 2–6, creamy with brown spots; I: 12–13 days; F: 10–12 days; B: 1–3.

Female, p. 41

Male

Tip Listen for the "chewink" call and rustling as it hops backward on ground, raking up leaf litter to find seeds.

Attracting

Attracted to seed scattered on the ground, such as sunflower hearts, millet, and cracked corn.

A large area of woods with sparse underbrush will attract them for nesting.

American Robin
Turdus migratorius 10"

 I.D. **MALE:** Dark gray above; orange-brown below; white under tail; bright yellow bill.

 Voice Song is a lively whistle, like "cheeryup cheerily"; calls include "teek" and "tuk tuk tuk."

 Habitat Lives in many habitats, from woods to open lawns, from plains to the timberline of mountains. **POPULATION:** ↑

 Nesting Nest of grass and mud, placed on limb of tree or building ledge. Eggs: 3–7, light blue; I: 12–14 days; F: 14–16 days; B: 2–3.

Female, p. 46

Male

Tip Most often seen running along lawns and tilting their heads to *see*, not hear, earthworms.

Attracting

 May come to feeders for fruit such as raisins or berries.

 Comes to birdbaths.

To attract, plant berry-producing shrubs and keep some areas of open lawn.

House Finch
Carpodacus mexicanus 5¹/₂"

Female, p. 28

Male

I.D.

MALE: Red on head and upper breast; broad brown streaking on lower breast and sides.

Voice

Both male and female give a song that is a musical warble ending with a harsh downslurred "jeeer."

Habitat

Urban areas, suburbs, parks, canyons, semidry brush country.
POPULATION: ↑

Nesting

Nest of twigs and grasses, placed in shrub, vine, hanging planter, or birdhouse. Eggs: 2–6, bluish white with speckles; I: 12–16 days; F: 11–19 days; B: 1–3.

Tip Common at feeders; often build their nests in hanging outdoor planters.

Attracting

Eats a variety of seeds, especially sunflower and thistle.

Hole: 1³/₈–2 in. dia. and 5–7 in. above floor
Floor: 4 x 4 in.

Comes to birdbaths to drink and bathe.

Purple Finch
Carpodacus purpureus 6"

I.D.

MALE: Upperparts, breast, and sides raspberry red; head uniformly red; little or no brown streaking on breast or sides.

Voice

Song is an extended warble; call while in flight is a short "pik."

Habitat

Mixed woods, coniferous forests, suburban yards. **POPULATION:** ↓

Nesting

Nest of twigs, grasses and rootlets, placed in tree. Eggs: 3–6, light blue-green with dark marks; I: 13 days; F: 14 days; B: 1–2.

Attracting

Attracted to sunflower, thistle, and millet seed, either scattered on the ground or in aboveground feeders.

Female, p. 29

Male

Tip Much less common at feeders than House Finch; distinguished from it by lack of brown streaks on sides.

Scarlet Tanager
Piranga olivacea 7"

I.D.
MALE: Scarlet-red body; black wings and tail. In winter, similar to female but with black wings.

Voice
Song is a well-spaced series of buzzy 2-part whistles, like "zureet zeeyeer zeeroo"; referred to as a "robin with a sore throat." Call is a distinctive "chip-burr."

Habitat
Mature deciduous forests.
POPULATION: ↑

Nesting
Nest of twigs, grasses, and rootlets, placed on horizontal limb of tree. Eggs: 2–5, pale bluish green with irregular dots of brown; I: 12–14 days; F: 9–10 days; B: 1.

Attracting
Comes to birdbaths for drinking and bathing.

Female, p. 8

Male

Tip Despite brilliant colors, male is hard to spot in tree leaves. Best discovered by song or "chip-burr" call.

18

Summer Tanager
Piranga rubra 7 1/2"

I.D.

MALE: Uniformly bright rose-red head and body; darker red wings and tail.

Voice

Song is a string of varied whistles; call is a rapid harsh "chchbit."

Habitat

Pine-oak woods, willows and cottonwoods along streams.
POPULATION: ↓

Nesting

Nest of weed stems, bark, and coarse grasses, placed on horizontal branch of tree. Eggs: 3–5, pale bluish green with dark marks; I: 12 days; F: unknown; B: unknown.

Male

Female, p. 9

Tip Like other tanagers, stays hidden in tree leaves most of time; best found by hearing song or call.

Attracting

Sometimes comes to feeders for fruit, bread crumbs, and suet mixtures.

Comes to birdbaths for drinking and bathing.

Since it is fond of berries, planting berry-producing shrubs will make a more favorable habitat for it.

19

Northern Cardinal
Cardinalis cardinalis 8 1/2"

I.D. **MALE:** All red with a crest; black on face around base of reddish bill.

Voice Female and male sing a series of clear whistles that vary, like "whoit whoit, cheer cheer cheer"; call is a metallic "chip."

Habitat Shrubs near open areas, woods, suburban yards. **POPULATION:** ↓

Nesting Nest of twigs, bark strips, and leaves, placed in dense shrubbery or small tree. Eggs: 2–5, buff-white with dark marks; I: 12–13 days; F: 9–12 days; B: 1–4.

Female, p. 42

Male

Tip At your feeder in spring, look for male feeding seeds to female as part of courtship.

Attracting

Readily comes to feeders. Prefers sunflower seed; also eats safflower seed, cracked corn, and mixed seed.

Comes to birdbaths for drinking and bathing.

Planting tall shrubs provides nesting habitats.

Common Yellowthroat
Geothlypis trichas 5"

I.D. **FEMALE:** Yellow throat and breast; whitish eye-ring; brown wash on forehead, back, and wings.

Voice Call is a sharp "tchet."

Habitat Dense brushy habitats near wet areas; also drier habitats with dense understory. **POPULATION:** ↓

Nesting Cuplike nest of grasses, leaves, and hair, placed in shrubbery. Eggs: 3–4, creamy white with brown marks; I: 12 days; F: 8–9 days; B: 1–2.

Attracting
Planting shrubs near a lake or pond may attract them to nest.

Male, p. 5

Female

Tip If you get near their nest, the birds will approach and give their "tchet" call.

Yellow-rumped Warbler
Dendroica coronata 5¹/₂"

Male, summer

Adult, winter

I.D.
In all seasons, note yellow rump and yellow patches in front of each wing. **FALL:** Brownish above; brown streaks on breast, sides, and back. **SPRING:** Female similar to fall but grayer. Male slate-gray above with yellow crown and black streaks on breast.

Voice
Song is a 2-note trill; common call is a "check."

Habitat
Summers in coniferous or mixed forests. In fall, brushy thickets.
POPULATION: ↑

Tip Look for these birds in fall, for they are very common during their southward migration.

Nesting
Cuplike nest of twigs, grasses, and rootlets, placed in conifer. Eggs: 4–5, cream with brown marks; I: 12–13 days; F: 12–14 days; B: 2.

Attracting
May come to feeders for suet and fruit.

Brown Creeper
Certhia americana 5 1/2"

I.D. Small brown bird that hitches up tree trunks; brown-streaked above; whitish below; relatively long downcurved bill.

Voice Song is a series of high-pitched whistles like "see wee see tu wee"; call is a high "tseee."

Habitat Woods. POPULATION: ↑

Nesting Hammocklike nest of bark and twigs, placed behind loose piece of bark on dead tree, or in natural cavity in tree. Eggs: 5–6, white with dark spots; I: 14–16 days; F: 13–15 days; B: unknown.

Tip Typically creeps up one tree and then flies to the base of the next tree before climbing up.

Attracting
Brown Creepers may come to suet, especially if the feeder is attached to a tree trunk where the bird normally might feed.

Indigo Bunting
Passerina cyanea 5 1/2"

I.D.
FEMALE: Brown overall, with faint wing bars and faint streaking; short, gray, conical bill.

Voice
Call is like a short "spit."

Habitat
Brush and low trees near open areas, such as overgrown fields.
POPULATION: ↓

Nesting
Nest of dead leaves, weed stems, and grasses, placed in fork of tree or on shrub branch. Eggs: 2–6, white; I: 12 days; F: 10–12 days; B: 1–2.

Attracting
Prefers millet and mixed seed scattered on the ground. May come to feeders during spring and fall migration and, in the far South, in winter.

Male, p. 64

Female

Tip Female is secretive. If you are near nest, she may fly to a perch and repeatedly call and flick her tail.

House Wren
Troglodytes aedon 5"

I.D.
A small bird with short tail often cocked up; upperparts grayish brown; underparts grayish white; some buffy barring on sides.

Voice
Song is a bubbling warble lasting 2–3 seconds, often repeated.

Habitat
Woods edges in rural or suburban areas; also mountain forests, clearings. POPULATION: ⇑

Nesting
Nest of short twigs lined with hair and rootlets, placed in any natural or human-made cavity, including birdhouse. Eggs: 5–6, white with brown marks; I: 12–15 days; F: 16–17 days; B: 1–2.

Tip Male builds nest foundations in several birdhouses; female selects one and adds final lining.

Attracting
Hole: 1–1½ in. dia. and 6–7 in. above floor
Floor: 4 x 4 in.

Carolina Wren
Thryothorus ludovicianus 6"

I.D.
Small bird with a short tail; warm brown above; rich buff below; prominent white eyebrow.

Voice
Song is a loud, repeated, 3-part phrase, like "tea kettle, tea kettle, tea kettle."

Habitat
Forest understory, vines, woodlands in rural or suburban areas.
POPULATION: ↑

Nesting
Nest of twigs, moss, and rootlets, placed in natural cavity, birdhouse, or other nook. Eggs: 4–8, creamy with brown marks; I: 12–14 days; F: 12–14 days; B: 2–3.

Attracting

Comes to suet feeders; also eats hulled sunflower seeds.

Tip Males can be heard singing in any month of the year.

Hole: 1 1/2 in. dia. and 6–7 in. above floor
Floor: 4 x 4 in.

Pine Siskin
Carduelis pinus 5"

I.D. Small brown-streaked bird with varying amounts of yellow showing on wings and base of tail; bill is fairly long and sharply pointed.

Voice Calls include a repeated "swee-yeet" and a distinctive, buzzy, ascending "zreeeee."

Habitat Coniferous or mixed woods, shrub thickets, suburban yards. POPULATION: ↓

Nesting Nest of grasses, twigs, and rootlets, placed in tree. Eggs: 1–5, light blue-green with dark marks; I: 13 days; F: 14–15 days; B: 1–2.

Attracting
Comes to hanging feeders with thistle, hulled or black oil sunflower seed. Will also feed on seed scattered on ground.

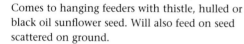
Tip Seen most often in winter at feeders, often in the company of goldfinches.

27

House Finch
Carpodacus mexicanus 5¹/₂"

I.D.

FEMALE: Uniformly finely streaked brown head; broad brown streaking on breast and belly.

Voice

Both male and female give a song that is a musical warble ending with a harsh downslurred "jeeer."

Habitat

Urban areas, suburbs, parks, canyons, semidry brush country. **POPULATION:** ↑

Nesting

Nest of twigs and grasses, placed in shrub, vine, hanging planter, or birdhouse. Eggs: 2–6, bluish white with speckles; I: 12–16 days; F: 11–19 days; B: 1–3.

Male, p. 16

Female

Tip Originally from West, expanding range in East; often build nests in hanging outdoor planters.

Attracting

Eats a variety of seeds, especially sunflower and thistle.

Hole: 1³/₈–2 in. dia. and 5–7 in. above floor
Floor: 4 x 4 in.

Comes to birdbaths to drink and bathe.

28

Purple Finch
Carpodacus purpureus 6"

I.D.

FEMALE: Well-defined pattern on face of a broad white eyebrow, brown eyeline, and white cheek. Broad, blurry, brown streaking on breast and belly.

Voice

Call while in flight is a short "pik."

Habitat

Mixed woods, coniferous forests, suburban yards. **POPULATION:** ↓

Nesting

Nest of twigs, grasses, and rootlets, placed in tree. Eggs: 3–6, light blue-green with dark marks; I: 13 days; F: 14 days; B: 1–2.

Attracting

Attracted to sunflower, thistle, and millet seed, either scattered on the ground or in above-ground feeders.

Male, p. 17

Female

Tip Much less common at feeders than the similar-looking female House Finch, which lacks white eyebrow.

House Sparrow
Passer domesticus 6"

I.D. **MALE:** Black bib; gray crown and cheek; rich brown back and nape. In fall, black bib is hidden by gray tips of fresh feathers. **FEMALE:** Grayish-brown breast; brown crown; buffy eyebrow; yellow bill.

Voice Calls include a "chirup chireep chirup."

Habitat Urban areas and parks; also farmland where livestock is present. **POPULATION:** ⇓

Nesting Messy nest of string, grass, and cloth, placed in crevice or birdhouse. Eggs: 3–7, white, or light blue with dark marks; I: 10–14 days; F: 14–17 days; B: 2–3.

Attracting
Eats all types of seed from a variety of feeders.

Male

Female

Male, fall

Tip Also called English Sparrow because it was introduced from England; most common sparrowlike bird in cities.

 Aggressive at nest sites and may kill or evict native species from birdhouses. Cannot enter nest holes smaller than 1 1/8 in.

Chipping Sparrow
Spizella passerina 5¹/₂"

I.D.

SUMMER: Clear gray breast; bright reddish-brown crown; thin black eyeline. **WINTER:** Head is buffier and less distinctly colored; brown crown has fine black streaks; eyeline is faint.

Voice

Song is a continuous rapid trill, 2–3 seconds long.

Habitat

Grassy areas, open woods, lawns, parks. **POPULATION:** ↑

Nesting

Cuplike nest of grasses and sometimes horsehair, placed on branch of tree. Eggs: 3–4, pale blue with dark blotches; I: 11–12 days; F: 7–10 days; B: 2.

Tip Common in suburban yards, nesting in evergreen foundation plantings.

Attracting

Comes to feeders with mixed seed or hulled sunflower seed scattered on the ground.

Planting dense evergreens in your yard is a good way to attract nesting Chipping Sparrows.

American Tree Sparrow
Spizella arborea 6"

I.D. Colorful sparrow with reddish-brown crown and eyeline; clear gray breast with black central dot; warm brown wings with white wing bars.

Voice Most common call is a 3-note "tseedle-eet."

Habitat Summers in subarctic scrub; winters in weedy fields, brushy edges, open woodlands, gardens.
POPULATION: ⇓

Nesting Cuplike nest of grasses and bark strips, placed on ground or in shrub. Eggs: 3–5, pale greenish or bluish white; I: 12–13 days; F: 9–10 days; B: 1.

Attracting
Most often comes to seed scattered on the ground or on a tray. Prefers cracked corn, millet, and hulled sunflower seed.

Tip Most often seen in winter in small flocks at feeders or near brushy edges.

Song Sparrow
Melospiza melodia 6"

I.D. Brown-streaked whitish breast with a dark central dot; long tail rounded at tip; gray eyebrow; heavy brown marks off base of bill.

Voice Song is repeated notes followed by a rich and varied warble, like "maids, maids, maids, put on your tea kettle ettle ettle."

Habitat Dense shrubs at edge of fields, lawns, streams. **POPULATION:** ↓

Nesting Cuplike nest of grasses, placed on ground or in shrub. Eggs: 3–5, greenish white with dark marks; I: 12–13 days; F: 10 days; B: 2–3.

Tip When disturbed, it often flies to the top of a nearby shrub and gives its "tchup" call.

Attracting
Comes to feeders where seed such as cracked corn, millet, or hulled sunflower is scattered on ground.

White-throated Sparrow
Zonotrichia albicollis 6½"

I.D. Two dark crown stripes; light eyebrow with yellow mark at front; white throat; clear gray breast. There are 2 forms of this species: white form has white eyebrow; tan form has buffy eyebrow.

Voice Song is 2 long whistled notes followed by 3–4 higher, quavering notes, like the phrase "sweet sweet Canada Canada Canada."

Habitat Coniferous and mixed woods, brushy areas. **POPULATION:** ↓

Nesting Cuplike nest of grasses lined with hair, placed on ground under small tree. Eggs: 4–6, light blue-green with dark marks; I: 11–14 days; F: 7–12 days; B: 1–2.

White form

Tan form

Tip Seen most in winter, when it moves south from its primarily northern breeding area.

Attracting
Comes to sunflower seed, cracked corn, or millet scattered on the ground.

Fox Sparrow
Passerella iliaca 7"

I.D. Large sparrow with reddish-brown streaking on its gray head and back; whitish underparts boldly streaked with brown or reddish brown; irregular central dot on breast.

Voice Song is a short series of melodious whistles.

Habitat Deciduous or coniferous woods, brushy areas, woods edges.
POPULATION: ↑

Nesting Cuplike nest of grasses and leaves, placed on ground under small tree or shrub. Eggs: 4–6, light blue-green with darker marks; I: 11–14 days; F: 7–12 days; B: 1–2.

Tip Mostly seen in winter, feeding on ground. They typically rake backward with their feet to expose seeds.

Attracting

May come to feeders with cracked corn, millet, or hulled sunflower seed scattered on the ground. Prefers the cover of shrubs or brush piles near feeders.

Common Ground-Dove
Columbina passerina 7"

I.D. Small pigeonlike bird with a short tail; gray with a scalloped effect on the feathers of head and breast. **IN FLIGHT:** Note reddish-brown wing tips.

Voice An ascending "coo ah, coo ah."

Habitat Open areas at the edge of taller vegetation, rural and suburban. **POPULATION:** ⇓

Nesting Saucerlike nest of sticks, placed on ground on beach, in woods, or in cultivated field. Eggs: 2–3, white; I: 12–14 days; F: 11 days; B: 2–4.

Attracting
Comes to feeders where seed is scattered on the ground. Prefers millet, cracked corn, hulled sunflower seed.

Tip Quite tame. Spend most of their time on ground in pairs or small groups.

Cedar Waxwing
Bombycilla cedrorum 7"

I.D. Sleek and crested; brownish above; yellow on belly; red dots on tips of wings; yellow tip to tail.

Voice Most Cedar Waxwing calls are very high-pitched whistles; flight call sounds like "seeee seeee."

Habitat Open rural or suburban areas. POPULATION: ⇑

Nesting Nest of grasses, twigs, and moss, placed in fork or branch of tree. Eggs: 2–6, pale with dark marks; I: 12–16 days; F: 14–18 days; B: 1–2.

Tip Almost always in flocks and often seen flying about trees and shrubs with berries.

Attracting
Attract Cedar Waxwings by planting trees and shrubs that produce berries or small fruits, which they eat.

Blue Grosbeak
Guiraca caerulea 7"

I.D. **FEMALE:** Plain brown overall; dark wings and tail; 2 buffy-brown wing bars; large, gray, conical bill.

Voice Call is a squeaky "chink."

Habitat Open areas with some shrubbery, such as roadsides, hedgerows, farmlands, prairies. **POPULATION:** ⇧

Nesting Nest of rootlets, grasses, and twigs lined with finer materials, placed in shrub, vine tangle, or tree. Eggs: 2–5, pale blue; I: 11–12 days; F: 9–13 days; B: 2.

Attracting

Comes to seed scattered on the ground, such as sunflower, peanut hearts, and cracked corn. Occasionally comes to tray and platform feeders.

Male, p. 69
Female

Tip Often twitches and rapidly spreads its tail when alarmed. Feeds on the ground.

Rose-breasted Grosbeak

Pheucticus ludovicianus 8"

I.D.
FEMALE: Brown; large, conical, pale bill; white eyebrow; heavy streaking on whitish breast.

Male, p. 99

Voice
Song is a loud series of slurred whistles; call is a distinctive squeak that sounds like a sneaker on a gym floor.

Habitat
Deciduous woods, areas with mixed trees and shrubs.
POPULATION: ↓

Female

Nesting
Nest of twigs lined with grasses and horsehair, placed in tree. Eggs: 3–6, pale blue with irregular brown spots; I: 12–14 days; F: 9–12 days; B: 1–2.

Tip Females also give song, near nest or while feeding; may show up at feeders looking like an oversized sparrow.

Attracting
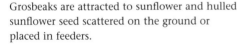
Grosbeaks are attracted to sunflower and hulled sunflower seed scattered on the ground or placed in feeders.

Comes to birdbaths for drinking and bathing.

39

Brown-headed Cowbird
Molothrus ater 7"

I.D.
FEMALE: Grayish brown overall with very little distinct marking; dark gray conical bill; faint streaking on breast. Often seen with more easily recognized male.

Voice
Calls include a high-pitched "pseeseee" and a chattering "ch'ch'ch'ch."

Habitat
Pastures, woods edges, lawns, forest clearings. **POPULATION:** ↓

Nesting
A female cowbird lays her eggs in the nests of other species, which then raise her young. Eggs: Usually only 1 per host nest, white with dark marks; I: 10–13 days; F: 9–11 days; B: unknown.

Attracting
Eats seed mixes scattered on the ground. Since cowbirds are parasitic on other birds, most people try to discourage them at feeders.

Male, p. 82

Female

Tip Females are often seen in the company of several males that are displaying as they compete for her.

Eastern Towhee
Pipilo erythrophthalmus 8"

I.D. **FEMALE:** Dark brown hood and back; sides reddish brown; belly white.

Voice Call is a loud "chewink."

Habitat Shrubby edges or open woods with shrub understory. **POPULATION:** ⇓

Nesting Nest of leaves, bark, and grass, placed on or near ground in scratched depression under brush. Eggs: 2–6, creamy with brown spots; I: 12–13 days; F: 10–12 days; B: 1–3.

Attracting

Attracted to seed scattered on the ground, such as sunflower hearts, millet, and cracked corn.

A large area of woods with sparse underbrush will attract them for nesting.

Male, p. 14
Female

Tip Listen for the "chewink" call and rustling as it hops backward on ground, raking up leaf litter to find seeds.

Northern Cardinal
Cardinalis cardinalis 8½"

I.D.
FEMALE: Buffy below; grayish brown above; reddish bill, crest, wings, and tail. Juvenile cardinal, seen in summer, looks similar but has a dark gray bill.

Voice
Female and male sing a series of clear whistles that vary, like "whoit whoit, cheer cheer cheer"; call is a metallic "chip."

Habitat
Shrubs near open areas, woods, suburban yards. **POPULATION:** ↓

Nesting
Nest of twigs, bark strips, and leaves, placed in dense shrubbery or small tree. Eggs: 2–5, buff-white with dark marks; I: 12–13 days; F: 9–12 days; B: 1–4.

Attracting

Comes to seed scattered on the ground or to hopper feeders. Prefers sunflower seed; also eats safflower seed, cracked corn, and mixed seed.

Female

Male, p. 20

Tip Often the first bird to visit your feeder in the morning and the last to visit at night.

Planting tall shrubs provides nesting habitats.

42

Red-winged Blackbird
Agelaius phoeniceus 8¹/₂"

I.D.

FEMALE: Brown above; heavily streaked brown below; sharp-pointed bill; buffy-to-whitish eyebrow.

Voice

A loud "ch'ch'ch'chee chee chee"; also "check" and "tseeert."

Habitat

Marshes and wet meadows.
POPULATION: ↓

Nesting

Nest of reeds and grasses attached to standing grass or shrub. Eggs: 3–5, pale greenish blue with dark marks; I: 11 days; F: 11 days; B: 2–3.

Attracting

Comes to feeders, especially in late summer, and eats seed scattered on the ground. Favors cracked corn and hulled or black oil sunflower seed.

Male, p. 85

Female

Tip Look for females among grasses and cattails, where they build the nests and raise the young.

43

Eastern Screech-Owl
Otus asio 9"

I.D. Small owl with yellow eyes and "ear tufts" (actually just feathers on top of its head, which can be lowered and hidden); pale bill. There are 2 forms: a reddish form and a gray form.

Voice An eerie whinny, rising and falling in pitch, and a long trill on one note.

Habitat Woods, swamps, deserts, parks, suburbs. POPULATION: ↓

Nesting Nests in tree cavity, old wood-pecker hole, or birdhouse. Eggs: 3–5, white; I: 21–28 days; F: 30–32 days; B: 1.

Tip **Often seen peering out of nest hole during the day. May catch nighttime insects at streetlights.**

Attracting
Hole: 3–4 in. dia. and 10–12 in. above floor
Floor: 8 x 8 in.

American Kestrel
Falco sparverius 9"

I.D. Small brownish falcon; two black sideburns on each side of face. **MALE:** Has blue-gray wings. **FEMALE:** Has reddish-brown wings.

Voice Common call is a series of sharp staccato notes like "klee klee klee klee" directed at intruders around nest.

Habitat A wide variety of open habitats, including urban areas. **POPULATION:** ↑

Nesting Nests in a natural cavity or birdhouse. Eggs: 3–7, pinkish with dark marks; I: 29–31 days; F: 29–31 days; B: 1.

Attracting

Kestrels can be attracted with nest boxes placed 15–30 ft. high on a pole or tree in open areas.

Female
Male

Tip Often seen perched atop trees along highways, where they hunt for voles.

Hole: 3 in. dia. and 10–12 in. above floor
Floor: 8 x 8 in.

American Robin
Turdus migratorius 10"

I.D. **FEMALE:** Brown above; pale reddish brown below; white under tail; bright yellow bill.

Voice Calls include "teek" and "tuk tuk tuk."

Habitat Lives in many habitats, from woods to open lawns, from plains to the timberline of mountains. **POPULATION:** ↑

Nesting Nest of grass and mud, placed on limb of tree or building ledge. Eggs: 3–7, light blue; I: 12–14 days; F: 14–16 days; B: 2–3.

Attracting

May come to feeders for fruit, such as raisins or berries.

Comes to birdbaths.

To attract, plant berry-producing shrubs and keep some areas of open lawn.

Male, p. 15

Female

Tip Most often seen running along lawns and tilting their heads to *see*, not hear, earthworms.

46

Northern Bobwhite
Colinus virginianus 10"

I.D. Small chickenlike bird with a short tail, light eyebrow and chin, and a wide dark streak through the eye. **MALE:** Has a white eyebrow and chin. **FEMALE:** Has a buffy eyebrow and chin.

Voice A variety of calls, including the whistled "bob-white" by the male and a "koilee" call given at dawn and dusk.

Habitat Farmland, brushy fields, open woods. **POPULATION:** ⇓

Nesting Nest is a scrape in the ground lined with grasses. Eggs: 12–14, white; I: 23 days; F: 14 days; B: 1.

Attracting
Eats cracked corn and other seed scattered on the ground; prefers cover and protection of brush pile near feeding area.

Female

Male

Tip Usually seen in small groups, running through grasses with their heads forward.

Likes dusty areas clear of vegetation for dust bathing.

47

Killdeer
Charadrius vociferus 10"

I.D. Dark brown above, whitish below; 2 dark neck-rings.

Voice Many varied calls. The most common is a repeated 2-part "killdeah," sounding much like the bird's name.

Habitat Open ground with gravel or short grass; suburban or rural.
POPULATION: ↓

Nesting Nest is a scrape in the ground with a few pebbles added. Eggs: 3–4, pale brown with darker marks; I: 24–28 days; F: 25 days; B: 1–2.

Tip Famous for their "broken-wing display," in which they look injured in order to distract predators from their nest or young.

Attracting
May nest in gravel areas or sparse lawns. If Killdeers nest in your area, try to keep dogs and cats away from their nest, which is placed on the ground.

Brown Thrasher
Toxostoma rufum 11"

I.D. Reddish-brown upperparts; heavily streaked breast and belly; long downcurved bill; long tail; pale eye.

Voice Song is a loud series of twice-repeated phrases imitating other birds' calls and songs.

Habitat Thickets and shrubs in open areas or at woods edges. POPULATION: ↓

Nesting Nest of twigs, grass, and grapevines, placed on ground, in bush, or in tree. Eggs: 2–6, light blue with fine dark marks; I: 12–14 days; F: 9–12 days; B: 2.

Tip Generally secretive, except in spring, when males sing from tops of shrubs and trees.

Attracting
Comes to birdbaths for drinking and bathing.

Plant large groups of dense shrubbery in open areas and you may attract nesting thrashers.

49

Sharp-shinned Hawk

Accipiter striatus 11"

I.D.
ADULT: Blue-gray above, lighter below with reddish-brown barring. IMMATURE: Brown above with brown streaking below.

Voice
A rapid series of high "kek kek kek kek" notes, most often given near nest.

Habitat
Summers in mixed deciduous and coniferous woods; winters in woods and near feeders.
POPULATION: ↑

Nesting
Nest is a platform of sticks lined with bark, placed in tree. Eggs: 3–8, whitish with dark marks; I: 30–32 days; F: 21–28 days; B: 1.

Attracting

Sharp-shinned Hawks sometimes catch birds at feeders; brush or cover near feeders provides safety for other birds.

Immature

Adult

Tip A small long-tailed hawk, usually first seen chasing small birds at feeders.

Mourning Dove
Zenaida macroura 12"

I.D. A sleek, gray-brown, pigeonlike bird; long pointed tail; large black dots on wings.

Voice Common call heard during spring and summer is a cooing that sounds like "ooahoo oo oo oo."

Habitat Can be found in almost any habitat. POPULATION: ↓

Nesting Nest is a loose platform of twigs, placed in tree. Eggs: 2, white; I: 14–15 days; F: 12–14 days; B: 2–3.

Attracting

Prefers seed scattered on the ground, such as white millet, sunflower seed, and cracked corn.

Comes to birdbaths, where it can suck up water through its bill, unlike other birds that need to tilt their heads back to swallow.

Tip During cooing call, given by unmated males, they puff out their throat feathers and bob their tails.

51

Northern Flicker

Colaptes auratus 13"

I.D. Red patch on back of neck; wide black necklace; whitish or buffy breast with black spots; brown-and-black-barred back and wings. IN FLIGHT: Note white rump and yellow underwings.

Voice A loud "kekekekeke" heard in early spring and a softer "woika-woikawoika" during courtship.

Habitat Parks, suburbs, farmland, woodlands. POPULATION: ⇓

Nesting Excavates nest cavity in dead tree, post, or cactus; may also use birdhouse. Eggs: 7–9, white; I: 11–12 days; F: 14–21 days; B: 1–2.

Attracting

Comes to suet at feeders; may also eat some seeds, such as hulled sunflower.

Male

Female

 Tip Flickers are often seen feeding on the ground, looking for one of their favorite foods — ants.

Hole: 2–3 in. dia. and 10–20 in. above floor
Floor: 7 x 7 in.

Boat-tailed Grackle

Quiscalus major 14"

I.D.

FEMALE: Buffy brown below; darker above with no iridescence. Eye color varies from yellow on Mid-Atlantic Coast to dark in Florida and on Gulf Coast.

Voice

Calls include a "chuck" and various squeaks.

Habitat

Salt marshes, parks, lakes.
POPULATION: ⇑

Nesting

Nest of grass and mud or cow dung lined with finer materials, placed in shrub or tree. Eggs: pale blue-gray with dark marks; I: 13–15 days; F: 20–23 days; B: 2–3.

Attracting

Comes to sunflower and mixed seed scattered on the ground.

Male, p. 87

Female

Tip Over the last 100 years, with the increase in irrigated farmland and lawns, their range has expanded westward.

Comes to birdbaths for drinking and bathing.

Male, p. 88

Female

Great-tailed Grackle
Quiscalus mexicanus 15"

I.D. **FEMALE:** Buffy brown below; darker above with slight greenish iridescence; yellow eye.

Voice Whistles, hisses, ratchety sounds, and clacks.

Habitat Open land with some trees; parks, urban areas. **POPULATION:** ⇑

Nesting Nest of grasses and mud or cow dung lined with finer material, placed in shrub or tree. Eggs: 3–4, bluish green with dark marks; I: 13–14 days; F: 20–23 days; B: 1–2.

Tip Nests colonially, sometimes in the thousands. Flies to huge roosts each night in winter.

Attracting

Comes to sunflower and mixed seed scattered on the ground. May try to feed at aboveground feeders, but usually has trouble holding on to the small perches.

Comes to birdbaths for drinking and bathing.

Wood Duck
Aix sponsa 18"

I.D. **FEMALE:** Grayish brown with a darker crown and broad white eye-ring that tapers to a point at the back.

Voice Call in flight is a distinctive "oo-eeek oo-eeek" given only by the female.

Habitat Wooded swamps and rivers. **POPULATION:** ⇑

Nesting Nest of wood chips and down, placed in natural tree cavity or birdhouse, over ground or over water. Eggs: 10–15, dull white; I: 27–30 days; F: 56–70 days; B: 1–2.

Attracting

A large birdhouse mounted on a pole a few feet above water can attract these ducks.

Hole: 3–4 in. dia. and 16–18 in. above floor
Floor: 10 x 10 in.

Male, p. 74

Female

Tip Female often seen taking care of the young by herself in summer.

Red-tailed Hawk
Buteo jamaicensis 19"

I.D. White chest; dark head; variable band of streaking across breast; tail reddish on adult when seen from above as the bird tilts and soars.

Voice A loud downslurred scream, like "tseeeaarr," sometimes given while soaring overhead.

Habitat Variety of open habitats, especially farmland, roadside grasslands. POPULATION: ⇑

Nesting Nest is a platform of twigs lined with bark and greenery, placed in tree. Eggs: 1–5, bluish white with dark marks; I: 28–35 days; F: 44–46 days; B: 1.

Tip Most common hawk seen perched in trees along highways, where it hunts voles in the grass.

Barred Owl
Strix varia 21"

I.D. Large owl; brown above, whitish below; dark eyes; no ear tufts; chest barred and belly streaked with brown.

Voice Hoots in phrase that sounds like "Who cooks for you?"— "hoo–hoo–hoohoo."

Habitat Woods and wooded swamps. POPULATION: ↑

Nesting Nests in tree cavity or abandoned hawk or crow nest. Eggs: 2–4, white; I: 28–33 days; F: 40–45 days; B: 1.

Tip Our only owl that regularly hoots during day or night.

Great Horned Owl
Bubo virginianus 22"

I.D. Very large owl; widely spaced ear tufts; yellow eyes; white throat that sometimes continues in a thin V down the chest.

Voice Four to six deep resonant hoots, given in various rhythms. Sometimes like "hoohoohoo hoohoo hoo."

Habitat Extremely varied; woods, suburbs, even deserts. **POPULATION:** ↑

Nesting Uses old nest of hawk or crow and adds feathers to the lining from its own breast. Eggs: 1–4, white; I: 28–30 days; F: 35 days; B: 1.

Tip Crows often mob this owl, and their drawn-out caws are often a clue to the owl's presence.

58

American Black Duck
Anas rubripes 23"

I.D. A large duck; dark brown body with a paler brown head and neck. **MALE:** Olive to yellowish unmarked bill. **FEMALE:** Greenish bill with gray mottling on top.

Voice Drawn-out "rhaeb" calls by the male; a variety of quacks by the female.

Habitat Summers on fresh- and saltwater marshes; winters along the coast. **POPULATION:** ↓

Nesting Nest of dry leaves and grasses lined with down from female's breast, placed on ground near water. Eggs: 6–12, cream or greenish buff; I: 26–29 days; F: 58–63 days; B: 1.

Female

Male

Tip Note their dark bodies and bill colors to distinguish them from female Mallard.

Attracting
Ponds and marshes with brushy edges are good nesting habitats for these birds.

Mallard

Anas platyrhynchos 24"

 I.D. **FEMALE:** A brown-streaked duck; orange bill broadly marked with black in the center; whitish tail feathers; dark blue feathers on wings sometimes visible.

 Voice Female gives a series of descending quacks; also a soft "quege-gege."

 Habitat Lakes, rivers, ocean bays, parks. **POPULATION:** ↑

 Nesting Nest of reeds and grasses lined with down, placed on ground near water. Eggs: 8–10, pale greenish white; I: 26–30 days; F: 50–60 days; B: 1.

Male, p. 75

Female

Tip Only females can make the quacking sound; male calls are a short whistle and a "rhaeb."

Attracting

Attracted to cracked corn scattered on the ground; will not come to feeders unless they are near water where ducks normally live.

Small ponds may attract them for feeding and/or nesting.

Ring-necked Pheasant
Phasianus colchicus 33"

I.D.
Chickenlike bird with a long pointed tail. **MALE:** Body is a mixture of iridescent greens, browns, and golds; has a white neck-ring, green head, and red wattles (bare skin on cheeks). **FEMALE:** Rich brown with darker markings on wings and back.

Voice
Male gives a "skwagock" call, and female responds with a "kia-kia" call.

Habitat
Farmland with woods edges and hedgerows. **POPULATION:** ↓

Female

Male

Tip Most often seen on ground. When flushed, flies up 10–20 ft. and then glides down.

Nesting
Nest of grasses and leaves, placed on ground in cover of grasses or shrubs. Eggs: 10–12, brownish olive; I: 24 days; F: 10–11 weeks; B: 1.

Attracting
Eats cracked corn and mixed seed off ground. Wary at feeders and prefers cover of tall grasses or shrubs nearby.

Canada Goose
Branta canadensis 36"

I.D. A large common goose seen in park ponds or grazing on lawns. Brownish gray overall; black head and neck; white chin.

Voice Male gives a low "ahonk" call; female gives a higher "hink" call, often alternating in a duet with the male.

Habitat Summers on lakes and marshes; winters on lakes, bays, fields, parks. **POPULATION:** ⇑

Nesting Nest of sticks, moss, and grass lined with down, placed on ground at edge of water or on grass hummock. Eggs: 4–7, white; I: 28 days; F: 2–3 weeks; B: 1.

Tip Common goose in parks. In long-distance flights, flies in V formations or diagonal lines.

Attracting
Eats cracked corn but in general should not be fed because it gets too used to humans and becomes a problem in public areas, fouling lawns and small ponds with droppings.

Brown Pelican
Pelecanus occidentalis 45"

I.D.
Large coastal bird; large dark bill; dark throat pouch; gray-brown body. Nape is dark brown in breeding adults; whitish in non-breeding adults. IMMATURE: All dark grayish brown.

Voice
Usually quiet when away from breeding grounds.

Habitat
Coastal.
POPULATION: ⇑

Nesting
Nests colonially on coastal islands. Nest is a rim of soil and debris on ground, or of sticks and grass in tree. Eggs: 2–4, white; I: 28–30 days; F: 71–88 days; B: 1.

Attracting
Attracted to fish in coastal bays and often roosts during the day in tall trees near these sites.

Immature
Adult, breeding

Tip Dives from air to catch fish in its pouch; small flocks seen gliding low over water in line formations.

Indigo Bunting
Passerina cyanea 5¹/₂"

I.D. **MALE:** Deep blue overall; short, dark gray, conical bill. In winter, similar to brown female.

Voice Song is a short rapid series of whistles, many paired, like "tsee tsee tew tew teer teer"; call is a short "spit."

Habitat Brush and low trees, overgrown fields. **POPULATION:** ↓

Nesting Nest of dead leaves, weed stems, and grasses, placed in fork of tree or on shrub branch. Eggs: 2–6, white; I: 12 days; F: 10–12 days; B: 1–2.

Attracting
Prefers millet scattered on ground. May come to feeders during spring and fall migrations and, in far South, in winter.

Female, p. 24

Male

Tip In spring and early summer, the male often perches at the tops of trees and sings.

Painted Bunting
Passerina ciris 5½"

I.D. **MALE:** Brilliantly colored; blue head; red underparts; light green back and wings.

Voice Song is a musical warble; call is a short "chit."

Habitat Brush, clearcuts, mesquite, rangeland, thickets. **POPULATION:** ⇓

Nesting Nest of weed stems, grasses, and leaves, placed in low foliage of tree or shrub. Eggs: 3–5, light bluish white or grayish white with brown spots; I: 11–12 days; F: 12–14 days; B: 2–4.

Attracting
May come to bird feeders for sunflower seed and seed mixes scattered on the ground.

Female, p. 73
Male

Tip Painted Buntings are secretive birds and often forage on the ground for seeds and insects.

Eastern Bluebird
Sialia sialis 6 1/2"

I.D. **MALE:** Brilliant blue head, back, wings, and tail; brick-red throat and breast. **FEMALE:** Light blue wings and tail; grayish-blue head and back; buffy throat and breast.

Voice Song is a series of downslurred whistles, like "cheer cheerful charmer"; call is a mellow "turwee."

Habitat Farmland, rural and suburban yards. **POPULATION:** ⇑

Nesting Nest of grasses and pine needles, placed in natural cavity or birdhouse. Eggs: 3–6, pale blue or white; I: 12–18 days; F: 16–21 days; B: 2–3.

Attracting

May come to feeders with suet, berries, or mealworms.

Male

Female

 Tip Populations have increased as people have put up more birdhouses.

Hole: 1 1/2 in. dia. and 7–8 in. above floor. Floor: 4 x 4 in.

 Comes to birdbaths for drinking and bathing.

66

Tree Swallow
Tachycineta bicolor 6"

Tip Generally seen in more rural areas, soaring through the air as they catch insects.

I.D. Iridescent blue above; pure white below. Females in their first 2 years are mostly brown above and white below.

Voice Common call is a loud "cheedeep cheedeep."

Habitat Open areas near woods and water. POPULATION: ⇑

Nesting Nest of grasses lined with feathers, placed in old woodpecker hole or birdhouse. Eggs: 5–6, white; I: 14–15 days; F: 21 days; B: 1–2.

Attracting

Tree Swallows readily accept birdhouses on poles out in the open.

Hole: 1¹/₂ in. dia. and 6–7 in. above floor
Floor: 5 x 5 in.

Barn Swallow
Hirundo rustica 7"

I.D. Upperparts iridescent blue; belly buff; throat reddish brown; tail forked.

Voice Both sexes give a song of continuous twittering interspersed by grating sounds; during feeding or alarm they call "chitchit."

Habitat Open country near barns or open outbuildings, bridges, culverts. POPULATION: ↑

Nesting Nest is deep bowl of mud pellets and grass lined with feathers, placed on beam or ledge in barn, under bridge, or in large culvert. Eggs: 2–7, white with reddish-brown speckles; I: 14–16 days; F: 18–23 days; B: 1–3.

Attracting
By creating ledges in an open outbuilding or leaving a window or door to a barn open, you may attract nesting Barn Swallows.

Tip Usually seen flying, it is best recognized by its long forked tail and reddish-brown throat.

Blue Grosbeak
Guiraca caerulea 7"

I.D.
MALE: Dark blue overall; 2 reddish-brown wing bars; black feathers around base of large silvery bill.

Voice
Song is a warbled phrase of musical notes; call is a squeaky "chink."

Habitat
Open areas with some shrubbery, such as roadsides, hedgerows, farmland, prairies. **POPULATION:** ⇑

Nesting
Nest of rootlets, grasses, and twigs lined with finer materials, placed in shrub, vine tangle, or tree. Eggs: 2–5, pale blue; I: 11–12 days; F: 9–13 days; B: 2.

Attracting

Comes to seed scattered on the ground, such as sunflower, peanut hearts, and cracked corn. Occasionally comes to tray and platform feeders.

Female, p. 38

Male

Tip Often twitches and rapidly spreads its tail when alarmed. Feeds on ground.

Blue Jay
Cyanocitta cristata 12"

I.D. Blue above; gray below; prominent crest; a black collar and necklace; wings and tail are spotted with white.

Voice A harsh "jaay jaay" given in alarm, a liquid "toolool," and a grating ratchetlike call. Also imitates the calls of hawks.

Habitat Woods and suburbs.
POPULATION: ↓

Nesting Nest of twigs, bark, and leaves, placed in tree. Eggs: 4–5, greenish blue, spotted with brown; I: 17 days; F: 17–19 days; B: 1–2.

Tip Roam in small flocks in fall and winter, looking for acorns, which they both eat and store.

Attracting

Eats a variety of seeds at feeders, including mixed seed, sunflower, and cracked corn. Often fills up its crop with sunflower seeds and goes off to store them in the woods, then returns for more.

Comes to birdbaths for drinking and bathing.

70

Belted Kingfisher
Ceryle alcyon 13"

I.D.
Large bill; grayish-blue head, back, and wings; white collar. **MALE:** All-white belly. **FEMALE:** Reddish-brown belly band.

Voice
A continuous woody rattle, often given in flight.

Habitat
Near water, such as rivers, lakes, coastal bays. Population: ↓

Nesting
Nest is a tunnel 3–15 ft. long with a chamber at the end, excavated in a vertical bank 1–2 ft. from the top. Eggs: 5–7, white; I: 22–26 days; F: 18–28 days; B: 1.

Female

Tip They often hover over water before diving down to catch a fish.

Ruby-throated Hummingbird
Archilochus colubris 3½"

I.D. **MALE:** Green upperparts; iridescent red throat; white breast and belly. **FEMALE:** Similar, but throat is white, not red.

Voice Sounds include various high-pitched chips and twitterings.

Habitat Woods edges, streams, parks, gardens. **POPULATION:** ⇑

Nesting Nest of plant down, bud scales, and lichens bound together with spider silk, placed on small horizontal limb of tree. Eggs: 2, white; I: 16 days; F: 30 days; B: 1–2.

Male

Female

Tip Hummingbirds feed 5–8 times per hour and for 30–60 seconds at a time; they are easy to miss seeing.

Attracting
Attracted to hummingbird feeders that contain a clear sugar solution.

Plant a hummingbird garden full of flowers with red tubular blossoms.

72

Painted Bunting
Passerina ciris 5¹/₂"

I.D. **FEMALE:** Leaf-green above; lighter green below; gray conical bill.

Voice Call is a short "chit."

Brush, clearcuts, mesquite, range-land, thickets. **POPULATION:** ⇓

Habitat

Nesting Nest of weed stems, grasses, and leaves, placed in low foliage of tree or shrub. Eggs: 3–5, light bluish white or grayish white with brown spots; I: 11–12 days; F: 12–14 days; B: 2–4.

Attracting
May come to bird feeders for sunflower seed and seed mixes scattered on the ground.

Male, p. 65

Female

Tip Painted Buntings are secretive birds and often forage on the ground for seeds and insects.

73

Wood Duck
Aix sponsa 18"

I.D.

MALE: Distinctive colorful head with iridescent green. White throat, partial neck-ring, and chinstrap.

Voice

Male gives high whistle; female gives distinctive "oo-eeek oo-eeek" call in flight.

Habitat

Wooded swamps and rivers. **POPULATION:** ⇑

Nesting

Nest of wood chips and down, placed in natural tree cavity or birdhouse, over ground or over water. Eggs: 10–15, dull white; I: 27–30 days; F: 56–70 days; B: 1–2.

Attracting

A large birdhouse mounted on a pole a few feet above water can attract these ducks.

Hole: 3–4 in. dia. and 16–18 in. above floor
Floor: 10 x 10 in.

Female, p. 55

Male

Tip Look for males gathering together in fall to do courtship displays that include flicking their chins up or to the side and raising wings and tails.

Mallard
Anas platyrhynchos 24"

I.D.
MALE: Iridescent green head; yellow bill; chestnut breast; white neck-ring may be hidden.

Voice
Male does not quack but gives a short whistle and a drawn-out "rhaeb" or short "rheb rheb" call.

Habitat
Lakes, rivers, ocean bays, parks.
POPULATION: ↑

Nesting
Nest of reeds and grasses lined with down, placed on ground near water. Eggs: 8–10, pale greenish white; I: 26–30 days; F: 50–60 days; B: 1.

Female, p. 60

Male

Tip Watch for males doing courtship displays in fall. They include the whistle call and fast swimming with head low.

Attracting

Attracted to cracked corn scattered on the ground; will not come to feeders unless they are near water where ducks normally live.

Small ponds may attract them for feeding and/or nesting.

75

I.D. White overall; short, stocky, yellow-orange bill. Breeding adult has buffy-orange plumes on its head, back, and breast.

Voice Gives a "kok" call when alarmed.

Habitat Open dry areas, lawns, fields, pastures with livestock.
POPULATION: ⇑

Nesting Nests with other herons in colonies. Nest of sticks, twigs, and reeds, placed in shrub or tree. Eggs: 2–6, bluish white; I: 21–24 days; F: 30 days; B: 1.

Tip A common small heron, often seen feeding along roads or in cattle fields.

Snowy Egret
Egretta thula 24"

I.D.
White body; black bill; black legs with bright yellow feet (its "galoshes"). During breeding, the feet turn orange or red.

Voice
During aggressive encounters with other herons, it gives a harsh "gaah" call.

Habitat
Coastal areas, marshes, river valleys, lake edges. **POPULATION:** ⇑

Nesting
Nests in large colonies or singly. Nest of sticks and twigs in a platform, placed on ground or in tree or shrub. Eggs: 3–5, pale bluish green; I: 20–29 days; F: 30 days; B: 1.

Tip Look for its yellow "galoshes" and black bill. May aggressively defend nest and feeding areas.

White Ibis
Eudocimus albus 25"

I.D. Large, white, long-legged bird; long, downcurved, reddish bill. **IN FLIGHT:** White wings show black tips.

Voice Mostly quiet. Alarm call is a nasal "hunk, hunk, hunk."

Habitat Salt and freshwater lakes, marshes, swamps, tidal mudflats, shores. **POPULATION:** ⇑

Nesting Nests in large colonies of thousands. Loose nest of sticks and twigs, placed in tree or shrub. Eggs: 3–5, greenish white with dark marks; I: 21–23 days; F: 28–35 days; B: 1.

Adult

Immature

Tip Ibises fly with their necks out straight, unlike herons and egrets, which keep them bent.

Little Blue Heron
Egretta caerulea 27"

I.D. **IMMATURE:** All white with greenish-yellow legs. Bill is two-toned — blue-gray at the base with a darker tip.

Voice Gives a low-pitched "aarh" during aggressive encounters.

Habitat Swamps, inland marshes, and coastal areas. **POPULATION:** ⇓

Nesting Nests colonially, often with other heron species. Nest of sticks and twigs, placed in tree or shrub. Eggs: 2–5, pale bluish green; I: 20–24 days; F: 42–49 days; B: 1.

Adult, p. 125

Immature

Tip This white plumage is kept 1 year, at which point the bird becomes mottled as it changes to the blue-gray adult plumage.

Great Egret
Ardea alba 39"

I.D. Large all-white heron; long yellow bill; legs and feet black. In the breeding season, adults grow long white plumes on their backs.

Voice Call is a deep rattlelike croak.

Habitat Marshes, swamps, seashores, lake margins. POPULATION: ⇑

Nesting Nests in colonies with other herons, ibises, and cormorants or singly. Nest is a flimsy platform of sticks, twigs, and reeds, placed in tree or shrub. Eggs: 1–6, pale bluish green; I: 23–26 days; F: 42–49 days; B: 1.

Tip **Feeds in water or on land by walking slowly with head forward and then striking prey.**

Chimney Swift
Chaetura pelagica 5¹/₂"

I.D. Always seen flying. Appears all black; has cigar-shaped body and sickle-shaped wings. Its fluttery wingbeats are rapid and done with stiffly held wings.

Voice Calls include a rapidly repeated "chitter-chitter" and a quick series of separate "chip" calls.

Habitat Rural or urban areas where there are chimneys in which they can nest. POPULATION: ↓

Nesting Nest of twigs held onto the inside of chimney with a sticky substance from the bird's mouth. Eggs: 4–5, white; I: 19 days; F: 14–18 days; B: 1.

Tip **Usually seen fluttering and soaring over buildings with chimneys as they catch insects.**

Attracting
If you have them entering a chimney, do not use the fireplace until the birds have left in late summer.

81

Brown-headed Cowbird

Molothrus ater 7"

I.D.

MALE: Glossy black body; dark brown head; dark gray conical bill.

Voice

Song is a liquid "bublucomsee"; calls include a high-pitched "pseeseee" and a chattering "ch'ch'ch'ch."

Habitat

Pastures, woods edges, urban lawns, forest clearings.
POPULATION: ↓

Nesting

A female cowbird lays her eggs in the nests of other species, which then raise her young. Eggs: Usually only 1 per host nest, white with dark marks; I: 10–13 days; F: 9–11 days; B: unknown.

Attracting

Eats seed scattered on the ground. Since cowbirds are parasitic on other birds, most people try to discourage them at feeders.

Female, p. 40

Male

Tip Usually seen in small groups, feeding on grassy areas in summer; seen in large flocks in fall and winter.

European Starling
Sturnus vulgaris 8"

Winter

Summer

I.D.

SUMMER: A common city bird. Glossy purple-black overall; long yellow bill. Juveniles, which form large flocks in late summer, are brown and gradually change to black. **WINTER:** Black body speckled overall with white and gold; black bill. White and gold spots wear off by spring.

Voice

Song is a stream of squeals, squawks, and imitations of other birds' calls.

Habitat

Cities and suburbs.
POPULATION: ↓

Tip Look for starlings perched near building nooks or tree holes, where they nest.

Nesting

Nest of grass, feathers, and flowers, in tree hole, birdhouse, or building crevice. Eggs: 2–8, light blue with dark marks; I: 12–14 days; F: 18–21 days; B: 1–3.

Attracting
Comes to feeders for suet and mixed seed.

Starlings, an introduced species, aggressively compete with native birds for nest holes. A birdhouse entrance hole of 1 1/2 in. or smaller keeps them out.

Purple Martin
Progne subis 8"

I.D.

MALE: All dark purple with black wings and tail. **FEMALE:** Dull purple back, tail, and wings; gray chest; whitish belly.

Voice

Call is a "cher cher" given near the nest; song is pairs of notes followed by a grating sound.

Habitat

Open areas, often near water and buildings. **POPULATION:** ↑

Nesting

Nest of mud, sticks, leaves, placed in special birdhouse. Eggs: 5–6, white; I: 15–16 days; F: 27–35 days; B: 1.

Male

Female

Tip **Purple Martins can be attracted to special birdhouses placed in open areas near buildings.**

Attracting

Martins in the East nest colonially in apartment-like birdhouses. Boxes must be monitored during breeding to be sure that birds are safe and healthy.

Hole: 2–2 1/2 in. dia. and 1 in. above floor
Floor: 6 x 6 in.

Red-winged Blackbird

Agelaius phoeniceus 8½"

I.D.
MALE: All black with a red shoulder patch, which is bordered by yellow; the shoulder patch can be hidden.

Voice
Song is a loud "okaleee"; calls include "check" and "tseeert."

Habitat
Marshes and wet meadows.
POPULATION: ↓

Nesting
Nest of reeds and grasses attached to standing grass or shrub. Eggs: 3–5, pale greenish blue with dark marks; I: 11 days; F: 11 days; B: 2–3.

Attracting
Comes to feeders, especially in late summer, and eats seed scattered on the ground. Favors cracked corn and hulled sunflower seed, but also eats other types of mixed seed.

Female, p. 43

Male

Tip Look for males perched atop shrubs or cattails in marshes, singing and spreading their wings.

85

Common Grackle
Quiscalus quiscula 12"

I.D.
Black with iridescence on head, back, and belly; yellow eye. Female has shorter tail and less iridescence than male.

Voice
Song is a short series of harsh sounds ending in a squeak, like "grideleeek"; calls include a "chaack" and "chaaah."

Habitat
Open areas with some trees; city parks, urban yards, farmland.
POPULATION: ↓

Nesting
Nest of grass and mud, placed in shrub or tree. Eggs: 4–7, pale greenish brown with dark marks; I: 13–14 days; F: 12–16 days; B: 1.

Male

Tip A common bird, often seen near water. Male flies with tail folded lengthwise during breeding.

Attracting

Attracted to sunflower seed and seed mixes scattered on the ground or on trays.

Comes to birdbaths for drinking and bathing.

Boat-tailed Grackle

Quiscalus major 16"

I.D. **MALE:** Large; long-tailed; all black with a bluish-green gloss on its head and back. Eye color varies from yellow on Mid-Atlantic Coast to dark in Florida and on Gulf Coast.

Voice Song is a series of harsh high-pitched notes, like "jeeb jeeb jeeb"; calls include a "chuck" and various squeaks.

Habitat Salt marshes, parks, lakes.
POPULATION: ⇑

Nesting Nest of grass and mud or cow dung lined with finer materials, placed in shrub or tree. Eggs: pale blue-gray with dark marks; I: 13–15 days; F: 20–23 days; B: 2–3.

Attracting
Comes to seed scattered on ground.

Male

Female, p. 53

Tip Look for them perched in plain sight near water or feeding on the ground in parks.

Comes to birdbaths for drinking and bathing.

87

Great-tailed Grackle
Quiscalus mexicanus 18"

I.D. **MALE:** Large; long-tailed; all black with purplish gloss on its head and back; bright yellow eye.

Voice Whistles, hisses, ratchety sounds, and clacks.

Habitat Open land with some trees; parks, urban areas. **POPULATION:** ⇑

Nesting Nest of grasses and mud or cow dung lined with finer material, placed in shrub or tree. Eggs: 3–4, bluish green with dark marks; I: 13–14 days; F: 20–23 days; B: 1–2.

Male

Female, p. 54

Tip Nests colonially, sometimes in the thousands. Flies to huge roosts each night in winter.

Attracting

Comes to sunflower and seed mixes scattered on the ground. May try to feed at aboveground feeders, but usually has trouble holding on to the small perches.

Comes to birdbaths for drinking and bathing.

American Crow
Corvus brachyrhynchos 18"

I.D. The familiar, very large, all-black bird with a large black bill.

Voice Varied types of caws, from short to long and descending.

Habitat Seen in just about all habitats. POPULATION: ↑

Nesting Nest of twigs and sticks lined with bark and grass, placed in tree. Eggs: 4–5, bluish green with brown marks; I: 18 days; F: 28–35 days; B: 1–2.

Attracting

Comes to all kinds of seed at feeders; will eat suet if it can get it and is attracted to food scraps.

Comes to birdbaths to drink.

Tip **In fall and winter afternoons, crows fly in large numbers to nightly roosts.**

Double-crested Cormorant
Phalacrocorax auritus 33"

I.D. Large; long neck and long bill; all black with a broad orange throat pouch beneath its bill. IMMATURE: In their first year, young cormorants are dark brown with a light brown throat and belly.

Voice Usually quiet away from nesting areas.

Habitat Coasts, inland rivers, lakes. POPULATION: ⇑

Nesting Nests colonially. Platform nest of sticks and seaweed, placed in tree or on ground. Eggs: 2–7, pale blue; I: 24–29 days; F: 35–42 days; B: 1.

Adult

Immature

Tip Look for cormorants perched near water on rocks or dead branches, often holding wings out to dry.

Turkey Vulture
Cathartes aura 26"

I.D. Very large; black with a small, featherless, red head; trailing half of wings silvery. When soaring, its wings are held in a V and it often tilts side to side.

Voice Vultures are silent except at the nest, where they may give grunts and hisses.

Habitat Open country and dumps, occasionally roost in urban areas. POPULATION: ↑

Nesting Nest is a scrape on bare ground, in cave, stump, cliff ledge, or old building. Eggs: 1–3, dull white; I: 38–41 days; F: 70–80 days; B: 1.

Tip They are most often seen soaring lazily along ridges or over fields, looking for carrion to eat.

Downy Woodpecker
Picoides pubescens 6"

I.D.

FEMALE: White spots on black wings; white belly and back; no red on back of head. Distinguished from similar Hairy Woodpecker by bill, which is about half as long as its head.

Voice

A high-pitched "teek" call and a call that sounds like the whinny of a miniature horse.

Habitat

Woods, farmland, suburbs.
POPULATION: ↑

Nesting

Excavates a nest cavity in dead wood. Eggs: 4–5, white; I: 12 days; F: 21 days; B: 1–2.

Attracting

Especially attracted to suet feeders but may also feed on hulled sunflower seed.

Male, p. 100

Female

Tip Seen hitching up tree trunks or along branches. May drum on resonant surfaces in spring as part of courtship.

Hairy Woodpecker
Picoides villosus 9"

I.D.

FEMALE: White spots on black wings; white belly and back; no red on back of head. Distinguished from similar Downy Woodpecker by its bill, which is almost as long as its head.

Voice

Calls include a high-pitched "teek" and a "wickiwickiwicki" sound.

Habitat

Woods, farmland, suburbs.
POPULATION: ↑

Nesting

Excavates a nest cavity in live wood. Eggs: 4–6, white; I: 11–12 days; F: 28–30 days; B: 1.

Male, p. 103

Female

Tip **Less common than Downy Woodpecker. Usually seen hitching up tree trunks or along large branches in older woods.**

Attracting
Especially attracted to suet feeders, but may also feed on hulled sunflower seed or sunflower chips.

93

Eastern Kingbird
Tyrannus tyrannus 9"

I.D. Black upperparts; white underparts; conspicuous white tip to its black tail.

Voice Calls include a harsh "kitter kitter kitter" and a short "kt'zee."

Habitat Open areas with some trees or shrubs. POPULATION: ↓

Nesting A messy nest of weed stems, bark, string, and feathers, placed on horizontal branch of tree. Eggs: 3–4, white with dark marks; I: 14–16 days; F: 14–17 days; B: 1.

Attracting
A few trees planted in an open area may attract a kingbird as both a nesting spot and a perch from which it can fly out to catch insects.

Tip Kingbirds are very aggressive; you may see them dive-bomb crows or hawks that fly near their nest.

Osprey
Pandion haliaetus 24"

I.D.
White head with a wide black streak through the eye. Dark back, wings, and tail; white underparts.

Voice
Loud downslurred chirp call repeated faster and at a higher pitch as the bird becomes increasingly alarmed.

Habitat
Large lakes, rivers, coast.
POPULATION: ⇑

Nesting
Large nest of sticks, placed in tree, on cliff, or on human-made platform. Eggs: 2–4, whitish with reddish-brown blotches; I: 34–40 days; F: 49–56 days; B: 1.

Tip Ospreys hover over water and dive down to catch fish, which they carry off in their talons.

Attracting
Sturdy platforms on very strong tall poles placed near the coast may attract Ospreys to nest.

Great Black-backed Gull

Larus marinus 30"

I.D.
Large gull; black back and wings; white body. Bill is yellow and legs are pink. **IMMATURE:** Whitish head contrasts with checkered brown-and-white back and wings; wide dark band at tip of tail. Takes 4 years to reach adult plumage.

Voice
Common call is a low-pitched "cowp cowp cowp."

Habitat
Coastal and the Great Lakes. **POPULATION:** ↓

Nesting
Nests colonially. Nest is a mound of seaweed and other vegetation, placed on ground or on rock ledge. Eggs: 2–3, olive with darker marks; I: 27–28 days; F: 49–56 days; B: 1.

Adult, summer

Adult, summer

Immature, 1st year

Tip **Easily recognized by black back. Often dominates other gulls for food and perches.**

Bald Eagle
Halieetus leucocephalus 31"

I.D.

Very large dark bird; white head and tail. **IMMATURE:** For the first 4 years head is dark and its tail and underwings are blotched with white.

Voice

High chirps and piercing screams.

Habitat

Along coasts, lakes, large rivers. **POPULATION:** ⇑

Nesting

Nest is a massive platform of sticks and vegetation, placed on cliff ledge or in top of large tree. Eggs: 1–3, bluish white; I: 34–36 days; F: 10–12 weeks; B: 1.

Immature

Adult

Tip Look for eagles soaring or perched along coasts, rivers, and large lakes.

Common Loon
Gavia immer 32"

I.D. Large waterbird of northern lakes. Black-and-white checkered back; black head and bill; black-and-white barred neck-ring.

Voice A drawn-out wail, a short tremulous call, and a long undulating call.

Habitat Summers on lakes, where it breeds; winters along the coast. POPULATION: ⇑

Nesting Platform nest of water vegetation, placed on ground on an island or at water's edge. Eggs: 2, olive-brown; I: 29 days; F: 2–3 months; B: 1.

Tip Watch for loons in the quiet coves of northern lakes and listen for their haunting calls, often given at night.

Attracting

If you see loons on a lake in summer, be careful not to upset them by running motorboats in the area or getting too near them, for they are sensitive to disturbances.

Rose-breasted Grosbeak

Pheucticus ludovicianus 8"

I.D.
MALE: Black-and-white bird with a striking bright red triangle on its breast.

Voice
Call is a squeak; song is a series of whistled notes that sound like a "robin in a hurry."

Habitat
Deciduous woods, areas with mixed trees and shrubs.
POPULATION: ↓

Nesting
Nest of twigs lined with grasses and horsehair, placed in tree. Eggs: 3–6, pale blue with irregular brown spots; I: 12–14 days; F: 9–12 days; B: 1–2.

Attracting

Grosbeaks are attracted to sunflower and hulled sunflower seed scattered on the ground or in feeders.

Male

Female, p. 39

Tip Listen for its call, which sounds like a sneaker skid on a gym floor.

Comes to birdbaths for drinking and bathing.

Downy Woodpecker

Picoides pubescens 6"

I.D.

MALE: White spots on black wings; white belly and back; red on back of head. Distinguished from similar Hairy Woodpecker by its bill, which is about half as long as its head.

Voice

A high-pitched "teek" call and a call that sounds like the whinny of a miniature horse.

Habitat

Woods, farmland, suburbs.
POPULATION: ↑

Nesting

Excavates a nest cavity in dead wood. Eggs: 4–5, white; I: 12 days; F: 21 days; B: 1–2.

Attracting

Especially attracted to suet feeders but may also feed on hulled sunflower seed.

Female, p. 92

Male

Tip Usually seen hitching up tree trunks. May drum on resonant surfaces in spring as part of courtship.

100

Red-headed Woodpecker
Melanerpes erythrocephalus 8"

I.D.
Only woodpecker with an all-red head. Body and wings boldly patterned with black and white. Juvenal plumage is similar, but with a brown head and back. This plumage is kept until midwinter.

Voice
Call is a loud "kweeer." Drumming on resonant surfaces is in soft short bursts.

Habitat
Farms, woods, suburbs.
POPULATION: ↓

Tip In addition to pecking in wood, may catch insects in the air and forage on the ground.

Nesting
Excavates nest hole in tree or telephone pole; may also use birdhouse. Eggs: 4–5, white; I: 12–13 days; F: 30 days; B: 1.

Hole: 1³/₄–2³/₄ in. dia. and 10–14 in. above floor
Floor: 5 x 5 in.

Attracting
Comes to suet at feeders.

Red-bellied Woodpecker

Melanerpes carolinus 9"

I.D.
Black-and-white bars on wings, back, and tail; back of neck red; head and belly tan. FEMALE: Red patch only on back of neck. MALE: Red patch extends to the forehead.

Voice
Call is a harsh "churrr" or double "chiv chiv." Drumming on resonant surfaces is in short bursts only 1 second long.

Habitat
Woodlands, parks, suburbs. POPULATION: ↑

Nesting
Excavates nest in living tree; may use birdhouse. Eggs: 3–8, white; I: 12–14 days; F: 25–30 days; B: 2–3.

Attracting

Readily comes to feeders to get suet, fruit, or seeds, especially hulled sunflower seed.

Male

Female

Tip The loud "churrr" or "chiv chiv" call is a good way to locate this woodpecker.

Female, p. 93

Male

Hairy Woodpecker
Picoides villosus 9"

I.D.

MALE: White spots on black wings; white belly and back; red on back of head. Distinguished from similar Downy Woodpecker by its bill, which is almost as long as its head.

Voice

A "teek" call and a "wickiwicki-wicki" call given in courtship.

Habitat

Woods, farmland, suburbs.
POPULATION: ↑

Nesting

Excavates a nest cavity in live wood. Eggs: 4–6, white; I: 11–12 days; F: 28–30 days; B: 1.

Tip Less common than Downy Woodpecker. Also drums on resonant surfaces in spring as part of courtship.

Attracting

Especially attracted to suet feeders, but may also feed on hulled sunflower seed or sunflower chips.

Pileated Woodpecker
Dryocopus pileatus 18"

I.D. Very large woodpecker, as big as a crow. Mostly black with a large red crest; white stripe down the neck. **FEMALE:** Black line off base of bill. **MALE:** Red line off base of bill.

Voice Gives long volleys of loud "cuk cuk cuk" calls; drumming on resonant surfaces in spring is usually soft and trails off at the end.

Habitat Mature forests in rural or suburban areas. **POPULATION:** ↑

Nesting Nest cavity is excavated in dead wood. Eggs: 3–5, white; I: 15–16 days; F: 28 days; B: 1.

Attracting
In some locations these woodpeckers come to

Male

Female

Tip This very large woodpecker excavates huge gashes in trees as it looks for carpenter ants.

suet feeders. In other areas they are more shy of humans.

Black-capped Chickadee
Parus atricapillus 5"

I.D.
Black cap and bib; white cheek; gray back; variable buff on sides. Best told from similar Carolina Chickadee by range and song.

Voice
Two whistled notes, the 1st higher than the 2nd, like "fee-bee." Calls include "tseet" and "chickadeedeedee."

Habitat
Woods, farmland, suburbs.
POPULATION: ⇑

Nesting
Nest in tree hole or birdhouse includes moss. Eggs: 6–8, white with speckles; I: 11–12 days; F: 15–17 days; B: 1–2.

Tip In winter, chickadees stay in small flocks that defend a territory against other chickadees.

Attracting
Prefers sunflower seed and suet in aboveground feeders.

Hole: 1⅛–1½ in. dia. and 6–7 in. above floor. Floor: 4 x 4 in.

Comes to birdbaths for drinking and bathing.

Carolina Chickadee
Parus carolinensis 5"

I.D.
Black cap and bib; white cheek; gray back; variable buff on sides. Best distinguished from similar Black-capped Chickadee by range and song.

Voice
Song is 4 notes, the 1st and 3rd higher than the others, like "fee-bee feebay."

Habitat
Woods, farmland, suburbs.
POPULATION: ↓

Nesting
Nest in tree hole or birdhouse includes moss. Eggs: 6–8, white with speckles; I: 11–12 days; F: 13–17 days; B: 1–2.

Tip In spring, chickadee winter flocks break up into mated pairs, which defend separate breeding territories.

Attracting

Prefers sunflower and suet in aboveground feeders.

Hole: 1⅛–1½ in. dia. and 6–7 in. above floor. Floor: 4 x 4 in.

Comes to birdbaths for drinking and bathing.

106

Tufted Titmouse
Parus bicolor 5"

Black-crested form

I.D. A small crested bird; gray above; white below; buffy sides. Black-crested form lives in Texas.

Voice Song is downslurred whistle, like "peter peter peter." Calls include a scolding "jwee jwee jwee."

Habitat Woods and suburbs.
POPULATION: ↑

Nesting Nest of moss, hair, and leaves, placed in natural cavity or birdhouse. Eggs: 4–8, white with small brown dots; I: 13–14 days; F: 17–18 days; B: 1–2.

Tip In fall and winter, titmice are usually in small family groups of 4–6 birds.

Attracting

Comes to sunflower seed and peanuts in above-ground feeders; also comes to suet.

Hole: 1³/₈–1¹/₂ in. dia. and 6–7 in. above floor
Floor: 4 x 4 in.

 Comes to birdbaths for drinking and bathing.

Red-breasted Nuthatch
Sitta canadensis 4½"

I.D.
Dark crown and nape; white face; black eye-stripe; rust-colored breast. **MALE:** Black cap and richly colored underparts. **FEMALE:** Gray cap and lightly colored underparts.

Voice
Common call is a nasal "nyeep nyeep nyeep."

Habitat
Coniferous woods.
POPULATION: ⇑

Nesting
Nest of rootlets, grass, and moss, placed in excavated hole, birdhouse, or natural cavity. Eggs: 5–7, white or slightly pink with brown spots; I: 12 days; F: 16–21 days; B: 1–2.

Attracting
Prefers sunflower seed and suet feeders.

Male

Tip **Nuthatches are our only birds that can climb headfirst down tree trunks. This species is seen mostly in winter.**

Hole: 1⅛–1½ in. dia. and 6–7 in. above floor
Floor: 4 x 4 in.

White-breasted Nuthatch
Sitta carolinensis 6"

I.D.
Dark crown and nape; white face; gray back. In general, the crown is black on males and gray on females.

Voice
Common call is a nasal "ank," given singly when the birds are calmly feeding or in a rapid series when they are disturbed.

Habitat
Woods. POPULATION: ⇑

Nesting
Nest of twigs, bark, and fur, placed in natural cavity or birdhouse. Eggs: 3–10, white with dark marks; I: 12 days; F: 14 days; B: 1–2.

 Tip Nuthatches are our only birds that can climb headfirst down tree trunks.

Attracting
Prefers sunflower seed and suet mixtures.

Hole: 1 1/8–1 1/2 in. dia. and 6–7 in. above floor
Floor: 4 x 4 in.

American Goldfinch

Carduelis tristis 5"

I.D.

WINTER: Grayish or brownish with only a hint of yellow on face and body; black wings and tail.

Voice

Flight call is "perchicoree perchicoree."

Habitat

Open areas with shrubs and trees, farms, suburban yards, gardens.
POPULATION: ↓

Nesting

Nest of weed bark fastened with caterpillar webbing, placed in shrub or tree. Eggs: 3–7, light blue; I: 12–14 days; F: 11–15 days; B: 1–2.

Attracting

Prefers thistle or hulled sunflower seed in hanging feeders.

Male, summer, p. 3

Winter

Tip Can look like different birds in summer because they change from mostly grayish brown to yellow.

Comes to birdbaths for drinking and bathing.

Dark-eyed Junco
Junco hyemalis 6"

I.D.
Dark gray to brownish gray with a white belly and pale bill.

Voice
Calls include "tsip" and "zeet"; song is a short trill.

Habitat
Summers in woods, bogs, mountains above tree level; winters in woods edges, brush, suburban yards. POPULATION: ↓

Nesting
Cuplike nest of grasses and moss, placed in depression in ground near tall vegetation. Eggs: 3–6, gray or pale bluish with dark blotches; I: 12–13 days; F: 9–13 days; B: 1–2.

Attracting

Prefers seed scattered on the ground, such as millet, cracked corn, or hulled sunflower.

Uses brush piles for daytime safety and nighttime roosting.

Tip One of the most common winter visitors at suburban and rural feeders; usually in flocks.

111

Eastern Phoebe
Sayornis phoebe 7"

I.D. Gray-brown above; whitish below; noticeably bobs its tail whenever perched.

Voice Song is a hoarse rendition of its name, like "feebee"; call is "chirp."

Habitat Woods, farmlands, suburbs. **POPULATION:** ↑

Nesting Nest is a mounded cup of mud and moss, placed on ledge or under bridge. Eggs: 4–5, white; I: 16 days; F: 18 days; B: 2–3.

Attracting
Placing a small wooden ledge under the eave of a house or shed may attract a phoebe to nest. They use the same nest site year after year and build on top of the old nest.

Tip Often nests on porch ledges or under bridges; one of the earliest arrivals in spring.

112

Sanderling
Calidris alba 8"

I.D.

SUMMER: Upperparts reddish to orange-brown with darker brown streaking; belly white; legs black. **WINTER:** Pale gray above; white below; black legs; straight relatively short (for a shorebird) bill; small black patch on shoulder sometimes visible.

Voice

Flight call is a quiet "kip."

Habitat

Summers along arctic tundra; winters along sandy coasts. **POPULATION:** ⇓

Nesting

Nest a scraped depression on dry tundra. Eggs: 4, olive-green with dark marks; I: 24–31 days; F: 17 days; B: 1–2.

Winter

Tip Most often seen on the beach in winter, barely avoiding the edges of the waves as it feeds.

Gray Catbird
Dumetella carolinensis 9"

I.D.
All gray with a black cap. Also has a reddish-brown patch under the base of its tail, but this is hard to see.

Voice
For song, catbirds mimic the calls of other birds, but repeat each imitation only once, not 3 times like the mockingbird. Common call is a catlike "meeow."

Habitat
Shrubs, tangled thickets, woods edges; rural to suburban.
POPULATION: ↓

Nesting
Nest of twigs, leaves, and grapevine bark, placed in shrub, vine, or small tree. Eggs: 2–6, dark blue-green; I: 12–14 days; F: 10–13 days; B: 1–2.

Attracting
Comes to feeders for suet, seed, and fruit.

Tip Catbirds are named for their meow call, which is given at the slightest disturbance.

Plant shrubs at the edge of open areas to attract.

Northern Mockingbird
Mimus polyglottos 11"

I.D. Grayish above; whitish below; long tail. IN FLIGHT: Note the bold white patches on its wings and white on its outer tail feathers.

Voice Mimics other birds' songs and calls, repeating each 3 or more times.

Habitat Open areas with shrubs; gardens, parks. POPULATION: ↓

Nesting Nest of twigs and leaves, placed in shrub. Eggs: 2–6, blue-green with brown marks; I: 12–13 days; F: 10–13 days; B: 1–3.

Tip One of the few birds to sing in fall; may also sing on moonlit nights from spring through fall.

Attracting
May come to feeders for fruits such as raisins.

Planting shrubs with berries may attract the birds to nest in them in spring and eat from them all winter.

Pigeon (Rock Dove)
Columba livia 13"

I.D.
The Pigeon is familiar to all. Due to breeding by humans, it can range in color from all white to all black, with just about anything in between.

Voice
At nest sites, such as above air conditioners, you hear "k't'-cooo"; during courtship displays the male calls "oorook'tookoo."

Habitat
Cities, parks, bridges, steep cliffs.
POPULATION: ↑

Nesting
Saucerlike nest of roots, stems, and leaves, placed on building ledge, rafter, or beam under bridge. Eggs: 1–2, white; I: 18 days; F: 25–29 days; B: 2–5.

Attracting
Comes to birdbaths to drink.

Typical form

Variations

Tip Pigeons originally came from England, where they nest on steep cliffs; that is why they love tall buildings in cities.

Covering ledges with screening can prevent nesting on houses.

Scissor-tailed Flycatcher
Tyrannus forficatus 14"

I.D. Very long split tail; pale gray with a pinkish wash on sides.

Voice Call is a "kit" and, during display flights, a "kakweee kakweee."

Habitat Open areas with scattered trees. POPULATION: ↓

Nesting Nest is a loose collection of soft natural and human-made materials, placed on horizontal limb or telephone pole. Eggs: 4–5, white with dark marks; I: 12–15 days; F: 14–16 days; B: 1.

Tip Common in Texas; often seen perched on wire fences or telephone lines.

Greater Yellowlegs
Tringa melanoleuca 14"

I.D. Long yellow-to-orange legs; long, thin, dark bill; whitish marks on dark upperparts; lighter underparts. The Lesser Yellowlegs is very similar but smaller and with a proportionately shorter bill.

Voice Call is a descending series of 3–4 notes, like "tew tew tew." Lesser Yellowlegs has a 2-note call, "tew tew."

Habitat Summers on subarctic forest bogs; winters on coastal marshes, beaches. **POPULATION:** ↑

Nesting Nest is a depression in the ground. Eggs: 4, buff with dark marks; I: 23 days; F: 18–20 days; B: 1.

Greater

Lesser

Tip Seen mostly from late summer through winter. Look for their long bright yellow legs.

Willet

Catoptrophorus semipalmatus 15"

I.D. Long, straight, rather heavy bill; long grayish legs. **WINTER:** Plain gray-brown above; whitish below. **SUMMER:** Brown streaking on head and neck; brown barring on breast. **IN FLIGHT:** Note distinctive bold white wing-stripe on black wing.

Voice Calls include a shrill "pill will willet" and a "kip kip kip."

Habitat Summers on coastal marshes in East and prairie marshes in West; winters on coastal marshes, beaches, mudflats. **POPULATION:** ↓

Nesting Nest a shallow scrape in ground lined with dry grass, placed in open area near water. Eggs: 4, olive with dark marks; I: 22–29 days; F: unknown; B: 1.

Winter

Winter

Summer

Tip Seen by most people in fall and winter along the coast, where it feeds on beaches and in mudflats.

Common Tern
Sterna hirundo 15"

I.D. Long pointed wings; thin, pointed, orange-red bill; black cap; upperwings are gray with darker tips.

Voice Calls include a short "kip" and a harsh descending "keear."

Habitat Lakes and coast.
POPULATION: ↓

Nesting Nests in colonies. Nest is a scrape in the ground. Eggs: 3, buffy with spots; I: 21–27 days; F: 28 days; B: 1.

Tip Often seen diving into water to catch small fish.

Laughing Gull
Larus atricilla 17"

I.D.

SUMMER: Black hood; dark red bill; white crescents above and below eye. Back and wings are dark gray. **WINTER:** Similar, but head is white with a grayish wash on the nape; bill is black. **IMMATURE—1ST YEAR:** Gray back; gray breast; black bill; gray-brown half-hood. Takes 3 years to reach adult plumage.

Voice

Typical call sounds like a high-pitched laugh — "ha ha ha ha."

Habitat

Coastal; may wander slightly inland. **POPULATION:** ⇑

Nesting

Nests in colonies. Nest is made of grasses and sedges, placed on ground. Eggs: 3–4, brownish with dark marks; I: 19–22 days; F: 35–40 days; B: 1.

Adult, winter

Immature, 1st year

Adult, summer

Tip Their laughlike call is a good way to identify these birds. They sometimes feed on aerial insects.

121

Ring-billed Gull
Larus delawarensis 19"

I.D.

ADULT: Clear black ring just before tip of thin yellow bill; back and wings light gray.
IMMATURE—1ST YEAR: Pale dark-tipped bill; gray back; light brown mottling on head and breast. Takes 3 years to reach adult plumage.

Voice

"Hyoh hyoh" and other calls.

Habitat

Coasts, lakes, dumps, fields.
POPULATION:

Nesting

Usually nests in colonies. Nest of grasses, pebbles, and sticks, placed on ground. Eggs: 3, light brown with dark marks; I: 21 days; F: unknown; B: 1.

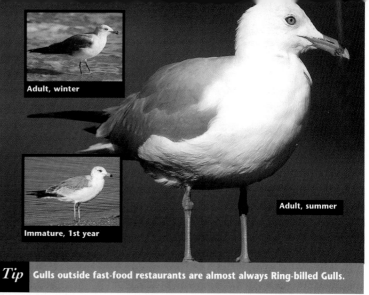

Adult, winter

Immature, 1st year

Adult, summer

Tip **Gulls outside fast-food restaurants are almost always Ring-billed Gulls.**

Herring Gull
Larus argentatus 25"

I.D.

ADULT: Wings and back light gray; head and body white; bill large, yellow, with a red spot near the tip of the lower bill; legs pinkish. **IMMATURE—1ST YEAR:** Uniformly mottled brown head and body; black bill with pale base. Takes 4 years to reach adult plumage.

Voice

Calls include the long-call, which sounds like "ow ow ow keekeekee kyow kyow kyow"; also gives a "huoh huoh huoh" call.

Habitat

Coasts, lakes, dumps, rivers, fields. **POPULATION:** ↓

Adult, winter

Immature, 1st year

Adult, summer

Tip Our most common large gull. Note red spot near tip of its lower bill.

Nesting

Nests in colonies. Nest is a scrape in ground lined with grasses and seaweed. Eggs: 3, brownish with dark marks; I: 26 days; F: 35 days; B: 1.

Tricolored Heron
Egretta tricolor 26"

I.D. Dark blue-gray heron with a white belly; white underwings seen in flight.

Voice High-pitched "aah" given during aggression.

Habitat Marshes, shores, mudflats, tidal creeks. **POPULATION:** ↑

Nesting Nests in large colonies, often with other herons. Platform nest of sticks lined with other vegetation, placed in tree or shrub. Eggs: 3–7, pale bluish green; I: 21–25 days; F: 35 days; B: 1.

Tip Its very long bill and white belly are the best ways to tell this from other grayish herons.

Little Blue Heron
Egretta caerulea 27"

I.D.

ADULT: Dark overall; purplish head; blue-gray body and wings; bill two-toned — blue-gray at the base with a darker tip. Immature is all white.

Voice

Gives a low-pitched "aarh" during aggressive encounters.

Habitat

Swamps, inland marshes, coastal areas. **POPULATION:** ⇓

Nesting

Nests colonially, often with other heron species. Nest of sticks and twigs, placed in tree or shrub. Eggs: 2–5, pale bluish green; I: 20–24 days; F: 42–49 days; B: 1.

Immature, p. 79

Adult

Tip Watch them dive down from the air to catch small fish, which they carry back to the young.

Great Blue Heron
Ardea herodias 50"

I.D. Very tall; grayish-blue body; white head; black stripe over eye.

Voice When competing with other herons, it may give a guttural "frahnk" or short "rok-rok."

Habitat Marshes, swamps, river and lake edges, tidal flats, mangroves, other water areas. **POPULATION:** ⇑

Nesting Nests in small colonies or singly. Nest is a large platform of sticks lined with other vegetation, placed in dead tree, often over water. Eggs: 3–7, pale bluish green; I: 28 days; F: 55–60 days; B: 1.

Tip Our largest and most widespread heron, it is usually seen feeding; often mistaken for a crane.